ancient environments

THE PRENTICE-HALL FOUNDATIONS OF EARTH SCIENCE SERIES

A. Lee McAlester, Editor

STRUCTURE OF THE EARTH

S. P. Clark, Jr.

EARTH MATERIALS

W. G. Ernst

THE SURFACE OF THE EARTH

A. L. Bloom

EARTH RESOURCES, 2nd ed.

B. J. Skinner

GEOLOGIC TIME, 2nd ed.

D. L. Eicher

ANCIENT ENVIRONMENTS, 2nd ed.

L. F. Laporte

THE HISTORY OF THE EARTH'S CRUST*

A. L. McAlester and D. L. Eicher

THE HISTORY OF LIFE, 2nd ed.

A. L. McAlester

OCEANS, 2nd ed.

K. K. Turekian

MAN AND THE OCEAN

B. J. Skinner and K. K. Turekian

ATMOSPHERES

R. M. Goody and J. C. G. Walker

WEATHER

L. J. Battan

THE SOLAR SYSTEM

J. A. Wood

*In preparation

ancient environments

second edition

LÉO F. LAPORTE

University of California, Santa Cruz

PRENTICE-HALL, INC., *Englewood Cliffs, New Jersey 07632*

Library of Congress Cataloging in Publication Data

LAPORTE, LÉO F (date)
Ancient environments.

(The Prentice-Hall foundations of earth science series)
Bibliography: p. 151
Includes index.
1. Paleocology—Addresses, essays, lectures.
I. Title
QE720.L36 1979 560 79-737
ISBN 0-13-036392-8
ISBN 0-13-036384-7 pbk.

Printed in the United States of America

10 9 8 7 6 5 4 3 2 1

PRENTICE-HALL INTERNATIONAL, INC., *London*
PRENTICE-HALL OF AUSTRALIA PTY. LIMITED, *Sydney*
PRENTICE-HALL OF CANADA, LTD., *Toronto*
PRENTICE-HALL OF INDIA PRIVATE LIMITED, *New Delhi*
PRENTICE-HALL OF JAPAN, INC., *Tokyo*
PRENTICE-HALL OF SOUTHEAST ASIA PTE. LTD., *Singapore*
WHITEHALL BOOKS LIMITED, *Wellington, New Zealand*

FOR MY FAMILY

contents

one

two

three

organisms and environments 49

four

taphonomy 83

five

environmental analysis 99

six

environmental synthesis 119

one

geologic environments

The history of organisms runs parallel with, is environmentally contained in, and continuously interacts with the physical history of the Earth. (George G. Simpson, 1963)

Scientific explanation is often expressed in "if . . ., then . . ." statements. That is, "if" certain necessary and sufficient conditions exist, "then" particular events will occur. For a simple example: If water is cooled to 0°C at one atmosphere of pressure, then it will undergo a change in state from liquid to solid. Similarly, the study of the Earth's environments is concerned with establishing if–then relationships by determining the necessary and sufficient conditions required for diverse geologic phenomena.

There is a broad range of geologic environments that demands inquiry or definition. What are the pressures and temperatures deep in the Earth's crust? What assemblages of silicate minerals will be at equilibrium under these temperatures and pressures? What are the states of stress and strain in active mountain belts or in stable continental blocks? What conditions of climate and habitat favored the invasion of land by the first amphibians? The investigation of these and many other geologic environments is being actively pursued by Earth scientists today.

Although research in recent and ancient geologic environments and their associated geologic processes has proliferated lately, the foundations of environmental analysis are as old as the science of geology itself.

One of the earliest controversies in geology, almost two centuries ago, concerned the geologic environment responsible for the formation of basalt, a dark, fine-grained rock composed of various silicate minerals. Abraham G. Werner, a German mineralogist of the eighteenth century, insisted that basalt, like all other rocks, was deposited from a "universal ocean" that once covered the earth. Werner denied that rocks could form in any way except as chemical precipitates from this universal ocean. Two of Werner's students, D'Aubuisson de Voissins and Leopold von Buch, influenced by the work of a French geologist, Nicholas Desmarest, realized that some rocks, basalt in particular, had an igneous origin—that is, they had crystallized with the cooling of a molten rock mass. For many years the young science of geology was divided into two bitterly opposed camps: the "Neptunists" who argued that all rocks were water-laid, and the "Plutonists" who maintained that certain rocks owed their origin to the eruption of hot masses of molten rock to the surface from below the Earth's crust. The issue, of course, was the correct geologic environment for different kinds of rocks found at the Earth's surface.

Modern interest in analyzing geologic environments has been spurred on by the development of new investigative tools, like the electron microscope; by the formulation of fresh theories, like plate tectonics; and by the extensive observations of active geologic processes as they occur naturally in the field or experimentally in the laboratory.

PALEOECOLOGY AND ENVIRONMENTAL STRATIGRAPHY

This volume considers one part of the broad spectrum of geologic environments, that of recent and ancient sedimentary environments and their associated organic remains. The goal of these investigations is to understand the complex interrelationships between ancient organisms and their habitats. This area of Earth science is called *paleoecology*. It is related to the field of *ecology*, which is concerned with explaining the interaction of *living* animals and plants with their physical, chemical, and biological environment. Ecology is an established science with its own body of data, concepts, and principles.

Ecology itself is subdivided into two areas: *synecology* and *autecology*. Synecology attempts to relate the abundance and distribution of *whole faunas and floras* to particular environmental regimes. Autecology seeks to explain the interactions of a *specific group* of organisms within the fauna and flora with local environmental conditions. For example, the synecology of organisms in a Pacific coral atoll describes which organisms feed on other organisms, how the animals and plants cope with the strong surf, and the role that sunlight and nutrients play in supporting the reef dwellers. The autecology of a particular species of sea urchin, on the other hand, will describe in what protected part of the reef the animal lives, how it feeds, and other relevant facts that distinguish it from other species in the atoll.

In one sense, then, paleoecology is simply ecology projected backward in time. Thus the paleoecologic study of a Paleozoic coral reef in Illinois will interpret the particular marine conditions that favored the proliferation of rugose and tabulate corals, pentamerid brachiopods, massive bryozoans, delicate crinoids, and robust trilobites; how the organisms interacted with each other and tolerated the physical and chemical factors present in the environment; and how individual species gathered food and occupied living space on the sea floor (Fig. 1-1).

But paleoecology, having a significant time dimension, can add another level of inquiry not usually found in most ecological studies. The analysis of how a given fauna and flora, whether a community of organisms in a coral reef, desert, or the deep sea, came to have the structure or organization it has. In other words, how the community evolved through geologic time. For example, has the basic ecologic structure of a modern coral reef been the same since the first appearance of coralline organisms, or have coral reefs evolved not only in terms of their constituent species but also in terms of how those species interacted with each other? It is this level of paleoecological inquiry, usually called *community evolution*, that intrigues many students and specialists today.

FIG. 1-1 This reconstruction of a Silurian coral reef community shows the inferred life habits of colonial tabulate corals (TC), solitary rugose corals (RC), crinoids (C), nautiloid cephalopods (N), trilobites (T), brachiopods (B), and snails (S) some 400 million years ago. Such a reconstruction represents graphically what the science of paleoecology seeks in part to describe and explain.

But there is another sense in which paleoecology is not merely ecology projected backward in time, owing to sparse preservation of many ancient organisms. Those organisms that are preserved as fossils are often extinct, so that we have no direct way of knowing what their vital needs were. Moreover, various environmental factors in the ancient habitat, such as temperature, salinity, and humidity, are not directly recorded in the existing sedimentary rock. Consequently, paleoecologists have been forced to develop their own techniques and procedures for inferring from the enclosing rock matrix what the original environmental conditions may have been, as well as for estimating numbers of individuals and kinds of organisms in the ancient environment. This leads us to the notion of *environmental stratigraphy.*

The intimate association of fossil organisms with sedimentary rocks demands that paleoecologists interpret the origin of the rock matrix as well as the presence of the included fossils. In fact, by using composition and size of sedimentary grains, primary structures, and other internal evidence contained in a sedimentary rock, paleoecologists can reconstruct the depositional environment of the rock. Such a reconstruction gives us information, independent of the fossils themselves, about the conditions under which the ancient organisms presumably lived and died or, at least, certainly were buried.

Because organisms have evolved throughout geologic history, we cannot uncritically interpret the past ecology of fossils by comparison to their living descendants. Even worse, many fossils are extinct without present-day close relatives. Fortunately, however, the kinds of rocks that form in a great variety of sedimentary environments have not become extinct. Hence, we can readily apply knowledge of recent environments to ancient rocks, and thereby define the boundary conditions, as it were, under which the fossils buried in the rocks probably lived.

ENVIRONMENTAL RECONSTRUCTION AND THE EARTH SCIENCES

Although it is intrinsically interesting to discover the kinds of environments in which various ancient fossil organisms flourished, paleoecology makes other important contributions to related geological fields of inquiry.

Paleoecology contributes most directly to *paleontology*, which is concerned with the history and evolution of life and therefore is a natural part of that science. For paleontologists seek more than merely a description of the various kinds of animals and plants that have lived in the past. They also wish to know why particular groups of organisms have evolved as they did and what environmental pressures the organisms were adapting to. To understand organic evolution, it is just as critical to know the habitats and habits of an organism as it is to know its shape and form.

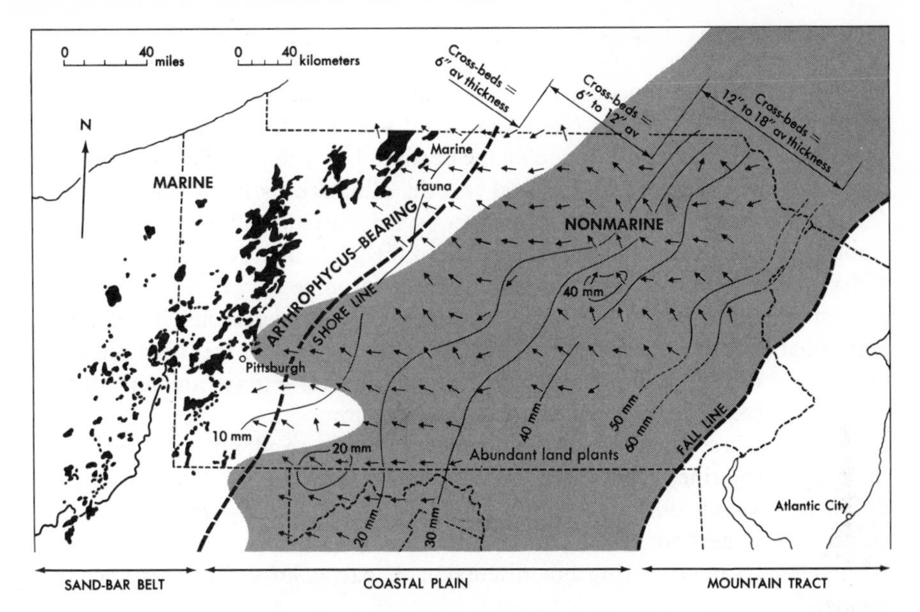

FIG. 1-2 Detailed analysis of the Pocono Formation, Mississippian Period, of the central Appalachians. This study by B. Pelletier included many aspects of this sedimentary rock, including grain size and composition, orientation of cross-stratification and plant remains, maximum size of quartz pebbles, sand/shale ratios, and fossil content. Pelletier was able to demonstrate that the Pocono Formation was a nonmarine, coastal-plain sediment derived from sedimentary rocks and low-grade metamorphic rocks in a source area located near Atlantic City, New Jersey. Sediment transport was to the west and northwest; the ancient shoreline trended northeast across Pennsylvania and was located some 25 miles east of Pittsburgh. Offshore the Pocono Formation is a marine shale and sandstone that contains abundant burrows ("Arthrophycus"), occasional brachiopods, and a few clams and snails. Sand/shale ratio greater than two is shaded; maximum pebble diameters in millimeters are shown by contours; current directions are shown by small arrows. Note relation of oil pools (black) to the sand bar belt. (After B. Pelletier, 1958.)

Paleoecology provides information not only regarding the distribution of ancient lands and seas, but also about what sorts of terrestrial and marine environments these might have been. Thus, besides indicating the position of former shorelines, paleoecology can help determine past climates, what kinds of landscapes existed where, and the prior location of ancient marine habitats like deltas, lagoons, and continental shelves. Such information is valuable to geologists attempting to describe and explain the long course of Earth history. In more practical ways, paleoecology contributes to the exploration for oil and gas trapped in marine sedimentary rocks. These fossil fuels accumulate from the decomposed remains of microscopic marine plants in certain types of porous and permeable rocks, often sand-bar deposits and coral reefs. Prediction of just where such deposits might occur depends on correct interpretation of the paleoecology and environmental stratigraphy of sedimentary strata lying deeply buried within the Earth's crust (Fig. 1-2).

ENVIRONMENTAL CLASSIFICATION

In defining and reconstructing ancient sedimentary environments and the paleoecology of the fossils that lived there, we have to consider the kinds of sedimentary environments found on the Earth today. The classification presented here is essentially based on physical criteria, although by implication chemical factors are involved as well. Moreover, since virtually all environments are populated by a variety of species adapted to the particular demands imposed by different environments, by extension, these environments will also presuppose certain biological factors. The classification of environments must be, of course, somewhat arbitrary, because the boundaries of one environment are not clear-cut from another. Nevertheless, certain environmental settings are quite distinctive, and these major categories are indicated below.

For each sedimentary environment it is useful to identify the *medium* of deposition (marine, freshwater, or subaerial), the *process* that deposits the sediments (waves, tides, rivers, wind, and so on), and the *place* of deposition (beach, deep-sea floor, tidal flat, lake bottom, desert, and so on). Keep these parameters in mind during our discussion of particular sedimentary environments.

Marine Environments

The environments that begin at the sea's edge can be broadly subdivided into two major realms: the *pelagic*, which refers to the water mass itself, and the *benthic*, which refers to the substrate of sediments at the bottom. The pelagic realm can in turn be subdivided into the water that lies over the continental shelves (*neritic* environment) and the water mass that lies beyond the continental shelves in the deeper ocean basins (*oceanic* environment). The oceanic water mass can be still further differentiated into various other subenvironments according to depth of water.

The benthic realm has subdivisions, too, which correlate more or less with the pelagic subdivisions. Thus, the *sublittoral* environment includes the sea bottom on the continental shelves, and hence is overlain by the neritic pelagic environment. The sea floor beyond the continental shelves includes the *bathyal*, *abyssal*, and *hadal* regions, which correspond more or less to the continental slope, the deep-ocean floor, and the deep-sea trenches, respectively.

That part of the sea floor that lies within the range of high and low tides is referred to as the *littoral* or *intertidal* environment. The narrow fringe of land that lies above the normal high water mark but is still within range of the sea's influence (salt spray, storm waves, or unusual high tides) is defined as the *supralittoral* environment. These various marine environments are illustrated in Fig. 1-3.

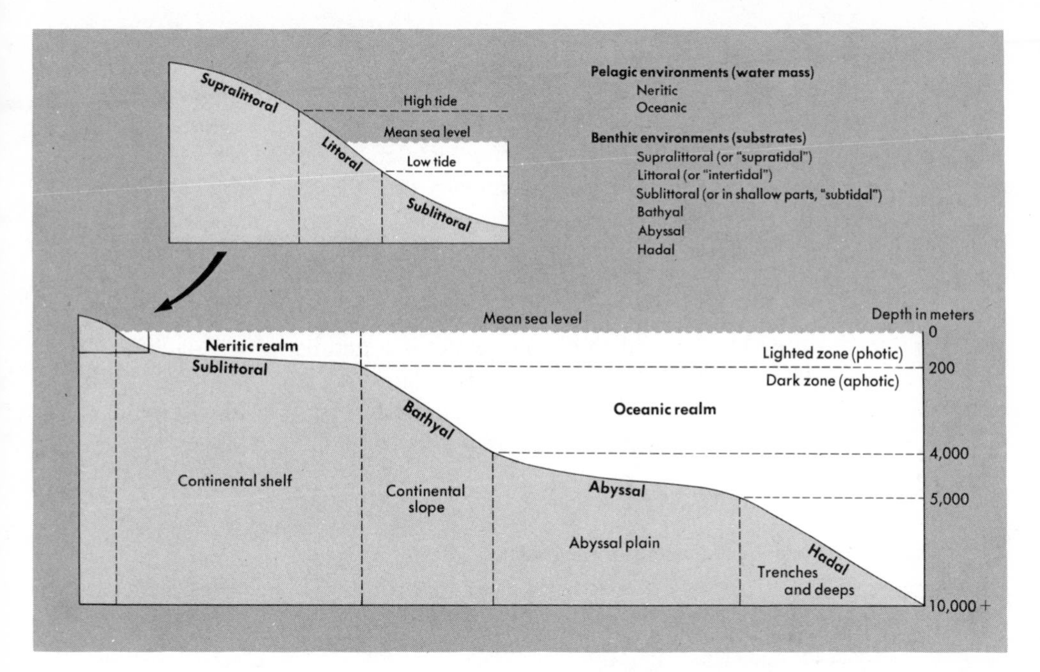

FIG. 1-3 Major marine environments as commonly defined. Although boundaries are not precise, note the general correspondence of sublittoral, bathyal, abyssal, and hadal environments with continental shelf, slope, abyssal plain, and trench. The penetration of sunlight decreases with water depth, being virtually absent in waters deeper than 200 meters, where it is always dark.

Terrestrial Environments

Terrestrial environments lying away from the sea's margin are considerably more diverse and individually more variable than marine environments. For example, except for very shallow and isolated bodies of the sea, marine environments do not have the great fluctuations of temperature that occur on land.

On the land surface itself there are various aquatic environments, such as the *lacustrine* (lakes and ponds) and the *fluvial* (streams and rivers) as well as *swamps* and *marshes*. Two particularly important ecologic factors in these aquatic environments are the amount of water movement and the ratio of water surface area to water depth. Water movement, or current, influences the circulation of oxygen and nutrients. Where the currents are especially strong, the local inhabitants may have special adaptations that enable them to move upstream or that prevent their being swept downstream and eventually even out to sea. The ratio of water surface area to water depth influences the amount and distribution of dissolved oxygen within the water mass. In shallow streams and ponds there is usually sufficient oxygen distributed throughout the water to support a rich flora and fauna. Deep lakes, by contrast, may have inadequate quantities of dissolved oxygen because of a relatively small water surface area in contact with the atmosphere and a slow rate of circulation of the water mass as a whole. Consequently, many deep lakes will have bottom waters so low in oxygen that few, if any, organisms will be able to live there.

In nonaquatic terrestrial environments, physical conditions such as temperature, humidity, wind, and sunlight fluctuate considerably not only during the year but even daily. These environments are, therefore, intrinsically far more variable than terrestrial aquatic environments and still more variable than marine environments where ecological conditions are generally more constant.

Besides these temporal variations in physical conditions, there are also rather rapid geographic differences related to changes in topography, latitude, proximity to the oceans, and so on. The result is a number of widely different dry-land habitats, including deserts, semiarid plateaus, arctic tundra, and rain forests. These various habitats can be characterized in terms of their dominant physical conditions together with the associated organisms adapted to these conditions.

Table 1-1 Major Environments of Deposition

- TERRESTRIAL
  - Subaerial
    - Landslide and talus
    - Dunes and desert pavement
  - Lacustrine
    - Lakes and ponds
    - Swamps
  - Fluvial
    - Alluvial fans
    - Rivers and streams
    - Flood plains
    - Deltas
- MARINE
  - Nearshore (subaerial to subaqueous)
    - Marshes
    - Dunes
    - Tidal flats
    - Beaches
    - Deltas
    - Lagoons
    - Estuaries
  - Offshore
    - Shallow subtidal (inner continental shelf)
    - Deep subtidal (outer continental shelf)
    - Continental slope
    - Deep sea
  - Organic buildups
    - Wave-built (e.g. shell mounds)
    - Organism-built (e.g. coral reefs)

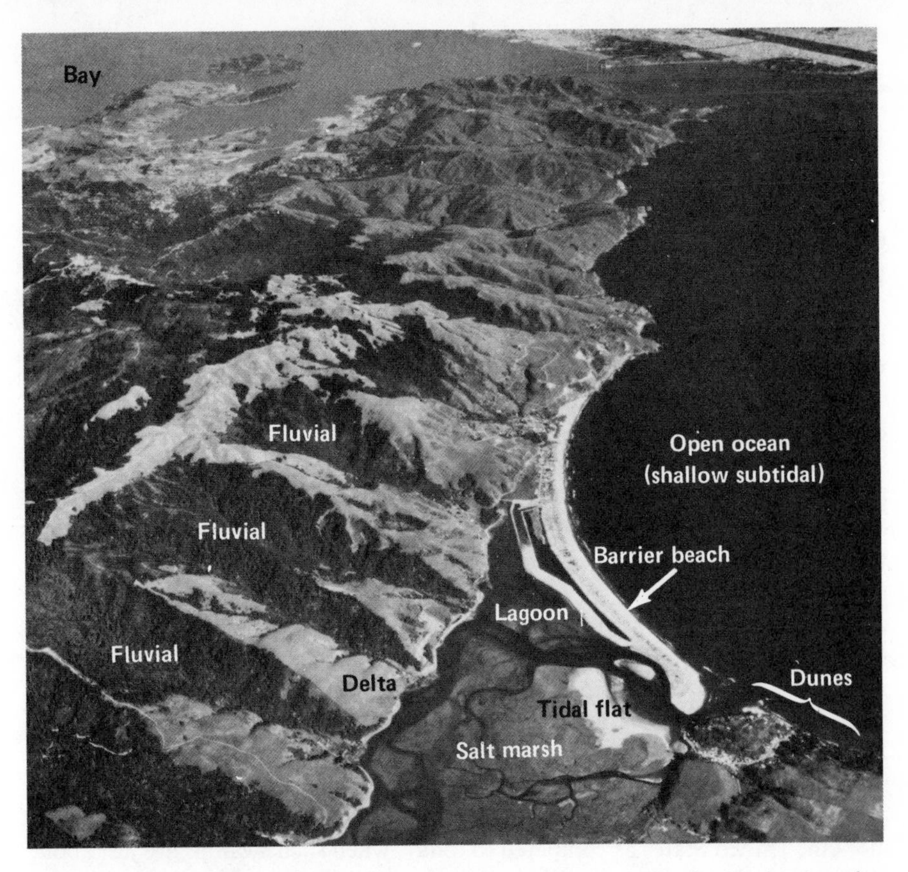

FIG. 1-4 Oblique aerial view of coast just north of San Francisco, California, showing the variety of environments from fully terrestrial to fully marine. Several of the major environments of deposition listed in Table 1–1 are identified.

Table 1-1 further subdivides marine and terrestrial environments into major different types. All of these present day depositional environments are recorded at different places and different times within the geologic record of ancient sedimentary rocks. Some of these environments are seen in Fig. 1-4.

ENVIRONMENTS THROUGH TIME

Paleoecology and the study of ancient environments are historical sciences, which means that we not only want to understand how physical, chemical, and biological processes operate in natural environments, but also the patterns, or configurations, that these processes have taken over the geologic past. As we will see, these configurations are recorded by the fossils and sedimentary rocks in the Earth's crust. Study of the progression of ancient environments over the several billion years of geologic time reveals three different kinds of environmental change: continually repeating, episodically repeating, and irreversible.

Continually repeating sedimentary environments refer to those shifting and migrating environments closely associated with the Earth's lands and seas.

Because the Earth is in constant flux, its mountains, rivers, lakes, shorelines and lagoons are forever altering their shape and geographic position. As the lands erode or are uplifted, as sea level rises and falls, the Earth's geography is constantly altered. The configuration of natural environments across the face of the Earth at this moment is, of course, but an instant in time. Over the next thousands and millions of years this configuration will slowly change. In those parts of the Earth's crust that are subsiding, sediments and fossilizable organisms will accumulate and record, like time-lapse photography, the temporal unfolding of environmental change. Over time, these environments will reoccur as the continents swell upward and erode, as the seas retreat and advance.

A second sort of environmental change through time is the irregular, episodic repetition of natural environments related to major, long-term crustal or global events. For example, we know that at various times in Earth history the crust was relatively inactive and the continents lay so low that in many places they were covered by very broad, shallow seas. For example, such epicontinental seas were widespread in late Cambrian, late Ordovician, and late Cretaceous times, judging from the extensive shallow marine rocks from these geologic periods found today on most of the world's continents. By contrast, such epicontinental seas are virtually absent today, the few examples being Hudson's Bay in Canada, the Gulf of Carpentaria in Australia, and the shallow shelf seas of Southeast Asia. Or consider the more extensive development of coastal swamps in the Carboniferous and Tertiary periods in eastern and intermontane United States as compared to today, when such swamps are just about restricted to the Florida Everglades and southern Louisiana. Global climates, too, have had similar, episodic fluctuations. Thus, widespread glaciation occurred in late Precambrian, Ordovician, late Paleozoic, and late Cenozoic times, whereas other periods of Earth history lack any such glacial record.

A third kind of environmental change that has happened over geologic history includes the irreversible events that have made the Earth a significantly different world. Such events are the origin of life and the evolution of the Earth's early atmosphere from one rich in carbon monoxide and dioxide, with lesser amounts of methane and ammonia, to one rich in nitrogen and oxygen. This slow oxidation of an early reducing atmosphere came about with the appearance and spread of photosynthesizing plants that later made possible the protective ozone layer in the upper atmosphere that shields out harmful solar ultraviolet radiation. Not until such an ozone layer formed could life leave water and colonize dry land. The continued evolution of life itself, of course, has effected other, relatively smaller, irreversible historical changes in the Earth's environments, at least in terms of their biology. A forest environment of the late Paleozoic Era with its seed ferns, scale trees, and primitive conifers that sheltered a rich and diverse amphibian and reptilian fauna is obviously a very different environment from modern North American forests.

These various kinds of environmental change through time have an important implication. As paleoecologists, we must consider what controls the distribution and abundance of animals and plants today in order to derive a body of principles and concepts that will guide our understanding of ancient biotic communities. But we must always remember that the study of the present cannot unequivocally establish past events and configurations important in evolutionary history. This is so because of the episodic and irreversible nature of some environmental changes. The present is a reasonable model for the past, but it is also one that ought to be used with intelligence and caution. The geological dictum that the "present is a key to the past" must be applied judiciously and with more than a grain of salt.

PLAN OF THE BOOK

The following two chapters deal with sediments and environments, and organisms and environments. In both we see how depositional environments leave their mark on the sediments and organisms occurring there. The next chapter explores *taphonomy*, or what happens to organisms from the time they die and are buried until the time they are discovered as fossils. The last two chapters consider how ancient environments are analyzed from the rock and fossil record, and then, once analyzed, how environmental data are synthesized into a coherent understanding of past biotic communities.

This volume, which emphasizes the environmental reconstruction of past events, is complementary to other volumes in this series, particularly *The History of Life*, by A. Lee McAlester, which treats the evolutionary development of ancient organisms, and *Geologic Time*, by Don L. Eicher, which considers the methods by which both ancient environments and fossil assemblages are dated within the sedimentary rock record.

SUMMARY

Several billion years ago, planet Earth coalesced from nebular dust and gas, slowly evolving into the physical world we know today of lands and seas, mountains and oceans, deserts and coral reefs. In tandem with this planetary evolution, life appeared and unfolded into all its richness and variety, occupying and flourishing in the myriad environments that we call Earth. Organisms past and present have adapted both as individual species and as communities of species to the physical, chemical, and biological conditions found in these environments.

Ecology is the study of the interactions of living animals and plants with their natural habitats. In a similar way paleoecology extends this study backward in time for fossil organisms. But whereas ecologists can directly record the significant environmental factors that control the distribution and abundance of organisms, paleoecologists must do this indirectly by studying the fossilized remains of life in sedimentary rocks. The reconstruction of ancient depositional environments depends upon environmental stratigraphy; the paleoecology of the included fossils may then be inferred.

Knowledge of present-day environments and ecological principles aids in our understanding of ancient environments and organisms. But this knowledge cannot be applied uncritically, for the Earth has experienced episodic and irreversible environmental changes over geologic time, so that what may be true today may not have been true throughout the geologic past.

two

sediments and environments

Behind the history of every sedimentary rock there lurks an ecosystem, but what one sees first is an environment of deposition. (Edward S. Deevey, 1965)

Sediments are deposits of solid material laid down by wind, ice, or water on the surface of the Earth. These deposits are as varied as beach sands, lake muds, stream gravels, coral reefs, and desert dunes. The two main sources of this variety are the origin of the sedimentary grains and the environment in which these grains are laid down. Another important variable in the study of sediments is what happens physically and chemically to them *after* deposition—through the processes of compaction, cementation, and recrystallization—so that they become rocks. Therefore, we must consider three separate environmental influences responsible for the formation of sedimentary rocks. These are (1) the genesis of the sedimentary grains in the source area; (2) the transportation and deposition of these grains in their final resting place; and (3) the transformation of the loose grains into a compact, lithified, sedimentary rock.

ORIGIN OF SEDIMENTARY GRAINS

Sediments initially form either within or outside of the area where they are ultimately deposited. Sediments may, for example, result from the erosion of pre-existing rocks that lie at various distances from the place where they will eventually accumulate. Thus, the Mississippi River annually deposits some half million metric tons of sediment that are derived from a very large region that includes all or part of 31 states, or 41 percent of the total area of the contiguous United States. Coast lines, too, are continually eroded by the surf along the sea's margins. The resulting erosional debris is deposited as local beaches and barrier bars, or it is carried along the coast by long-shore currents and laid down a considerable distance from its place of origin.

Sediments may also originate within the area of deposition. For example, shelly invertebrates extract calcium carbonate from sea water to build their skeletons; when the organisms die their shells are deposited with other accumulating sediments. In some hypersaline lakes and seas, salts such as sodium carbonate and calcium sulfate precipitate because of high rates of evaporation. There are various places in the world—such as Pakistan, Russia, and Germany, as well as Kansas and Michigan in the United States—where ancient salt deposits are profitably mined for table salt, gypsum, and potassium.

Sediments and rocks composed of grains that have been broken down and eroded as discrete particles—whether boulders, pebbles, sand, or silt—from the source area are called *clastic* (from Greek, meaning broken). *Nonclastic* sediments and rocks form as precipitates from saline lakes and sea water as well as from materials secreted by animals and plants. Thus we see that the composition of a particular sediment or rock will reflect the composition of the rocks being eroded in the source area as well as the nature of the inorganic and organic precipitates that might also be forming in the area of sediment deposition.

The composition of sediments also depends on the *rates* of both weathering in the source area and deposition in the sedimentary basin. If rocks in the source area are deeply weathered, then their constituent minerals are chemically altered and mechanically disintegrated. If erosional rates are rapid, then the minerals are transported and buried before much alteration and disintegration can occur.

Consider the weathering of granite, which is composed of quartz, mica, and feldspar. If the rates of weathering and transport are relatively slow, the micas and feldspar will have ample time to break down into fine-grained clay minerals; the quartz grains may be rounded, but they will not alter chemically because of the great stability of quartz. The resultant products of this granite therefore will be brought to the area of deposition as fine quartz sand mixed with finer-grained clays. The quartz sands may accumulate as nearshore or beach sands while the clays are carried farther offshore, where they eventually settle out of suspension as mud. If this same granite had been subjected to more rapid rates of weathering and transportation, then the sedimentary deposits would have been quite different. The feldspars and micas would have been incompletely weathered and thus little altered, and would be deposited as fine sand mixed with the quartz grains, yielding a sediment of different composition and texture from that in the first case (Fig. 2-1).

It should be noted in passing, too, that rates of erosion and transportation depend strongly on climate and topographic relief of the source area. For example, wet climates favor chemical alteration of source rocks, because water plays an important role in many chemical reactions. Rocks on steep slopes move more readily under the influence of gravity than rocks on more gentle slopes.

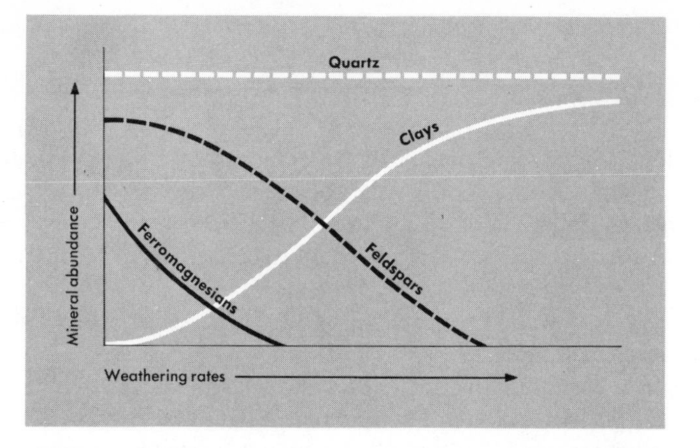

FIG. 2-1 Changes in the relative abundance of minerals with increasing weathering rates. For this example granitic mineral constituents are shown: *ferromagnesians* (mica and some amphibole), *feldspars*, and *quartz*. Clays are chemical alternation products of ferromagnesians and feldspars.

Not only are landslides and soil creep aided by steep slopes, but the velocity of flowing water is greater, thereby allowing it to move more surface debris. Refer to Table 2-1, which illustrates major kinds of sediments and sedimentary rocks according to origin of sedimentary grains, their textures, and how they are deposited. We will now discuss some of these other aspects of sediments and sedimentary rocks.

Table 2-1 Origin and Classification of Major Sedimentary Rock Types

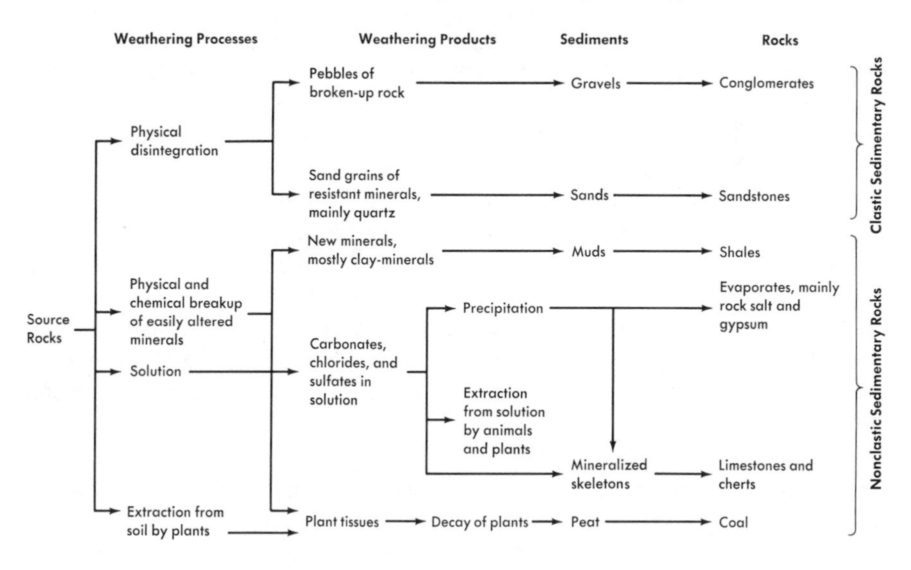

From Laporte, 1975

TRANSPORTATION AND DEPOSITION OF SEDIMENTS

Water is the principal agent of sediment transport. When water falls as rain it is, at first, quickly absorbed by the soil. Soon, however, the upper layers of soil become saturated and the rain water begins to run off across the ground's surface. This surface runoff forms small rivulets that join to form brooks, streams, and eventually rivers.

During its overland journey the flow of water transports sedimentary materials in three different ways. First, the water takes into *solution* various substances such as calcium, iron, and carbon dioxide. Second, fine-grained minerals and rock fragments are carried in *suspension* in the turbulent flow of the running water. Third, the water flow moves coarse-grained particles by *traction*, bouncing and rolling them along the stream or river bed.

Suspended and tractive sediments are eventually deposited in the delta at the river's mouth. Most of the dissolved load, however, goes directly into the sea, where it may either precipitate later (for example, by organisms as calcium carbonate) or where it may remain indefinitely in solution (for example, sodium chloride).

Ice, too, may be a significant agent of sediment transportation. For at various times in the past history of the Earth, thick masses of snow and ice covered large parts of the continents. As the snow collected and compacted, it turned to ice and under its own weight began to flow as a very viscous solid, expanding outward from its place of initial accumulation. As these thick glacial masses inched inexorably forward, they skimmed off soil and weathered rock layers. Some of this glacially eroded debris became frozen within the glacier only to be washed out, perhaps many miles distant from where it was first gathered up, when the glacier ice eventually melted. The hummocky topography in the northern latitudes of the Northern Hemisphere is the result of the deposition of sand and gravel by Pleistocene glaciers, which were widespread in these areas.

Wind is a much less dense and viscous medium than either water or ice and therefore usually carries far less sedimentary material in suspension or traction, and virtually none in solution (although vater vapor in the atmosphere may contain some dissolved salts). In areas where there is a poor cover of vegetation and where the climate is arid, there may be significant sediment transport by the wind, resulting in the formation of sand dunes. Windblown sand may also be a very effective erosional agent by abrading rock outcrops and desert pavements.

Transportation of sediments within the sea or along its margin is accomplished solely by moving sea water, although there may be occasional rafting of sediment out to sea by debris-laden icebergs. Movement of sedimentary grains within the oceans is basically the same as that in streams. Yet, although rivers and streams have confined channels along which sedimentary particles are transported, currents within the sea are often less well defined. For example, the Gulf Stream sweeps across the Blake Plateau, a particularly wide extension —more than 300 kilometers in some places—of the continental slope of the southeastern United States. Although the water depths over the Blake Plateau range from about 200 meters to more than 1,000 meters, the surface of the plateau is only thinly veneered by recent marine sediments; rocks of Tertiary age crop out at or near the surface. Marine geologists have inferred from this that the broad surface of the Blake Plateau is swept clean of any sedimentary material by the Gulf Stream whose axis of flow shifts periodically back and forth across the plateau.

Besides wind-induced and tidal currents, *turbidity currents* are also effective in the removal and transportation of marine sediments. Owing to the great topographic relief of the sea floors, intermittent movement of watery muds and sands by gravity flow occurs. Such turbid, sediment-laden currents, which are often triggered by earthquakes, can erode older, consolidated marine sediments. It has been suggested by several marine geologists that the lower parts of submarine canyons that cut across the continental shelves and slopes have been carved out by such turbidity flows as material is moved from the continental margins down into the abyssal plains of the deep ocean basins. The upper portions of these canyons were cut by rivers flowing across the continental shelves when Pleistocene glaciation lowered sea level.

There are often relatively long intervals during which sediments are not transported any significant distance in their journey from their source area to their final accumulation site. For the agents of sediment transportation vary in their capacity to carry materials and in their activity. For example, significant quantities of sediment may be transported only during the flood stage of a river. If so, between floods, just the dissolved fraction and the fine-grained, suspended fraction of a river's sediment load will move downstream (Fig. 2-2). Sediments

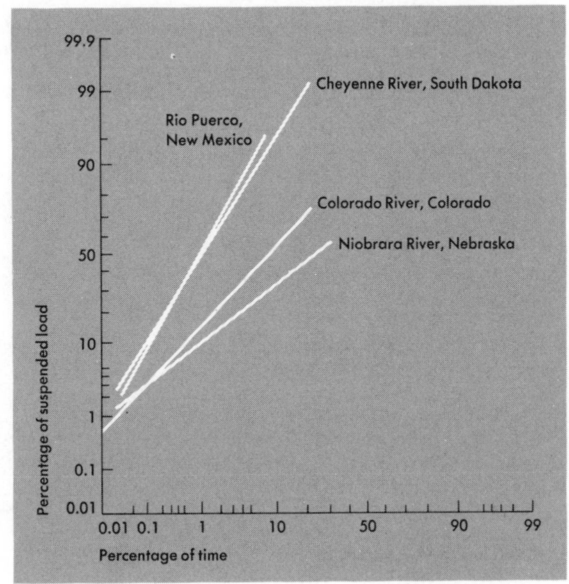

FIG. 2-2 Graphs of four rivers showing the times required to transport a given fraction of their suspended load. Note that most of the suspended sediment is carried only a short part of the year, owing to annual variations in discharge of water in the rivers, which, in turn, reflect seasonal fluctuations in precipitation. Fifty percent of the suspended load is carried only about four days of the year for the Rio Puerco and Cheyenne, 31 days for the Colorado, and 95 days for the Niobrara.

eroded in high, mountainous areas are deposited as alluvial fans within the adjacent valley floor where they can remain for long periods of time. But eventually, they, too, are eroded and retransported, gradually progressing toward the delta at their river's mouth. As for sediments that accumulate initially along the continental shelves, occasional turbidity currents will later transport some of this material farther out to sea in the deeper parts of the oceans. Therefore, although the net movement of sediment may be relatively slow by human standards, given the great eons of geologic time available for erosion, transportation, and deposition, the overall impact of these geologic processes is enormous.

Texture is a particularly useful characteristic in describing and interpreting a sedimentary rock. Texture refers not only to the size of the component grains, but also to their shape and mutual arrangement within the enclosing matrix. Sedimentary textures provide clues to the nature of the depositing medium. For example, the size and angularity of stream-laid deposits increase exponentially with an increase in the velocity of stream flow. Sediments that are coarse-grained, angular, and poorly sorted indicate rapid deposition by swift-moving water. On the other hand, sediments that are fine-grained, well-sorted, and laminated suggest deposition in quiet water, where individual small grains settle slowly out of suspension. Experimental data, shown in Fig. 2-3, support this qualitative observation of the relationship between water velocity and size of material eroded, transported, and deposited. For sand-sized and coarser grains, increasingly larger grains are eroded and transported as water velocity increases. For particles smaller than sand—silt and clay—water velocities either decrease if the fine materials behave as discrete grains, or increase if they cohere together, as clays often do, and behave as larger aggregates.

FIG. 2-3 Graph showing size of grains that will be eroded, transported, or deposited at a given velocity. Notice that *cohesive*, fine-grained sediments resist erosion by high-velocity water in the same way coarse sediments do. These experimental data reasonably approximate what is observed in nature.

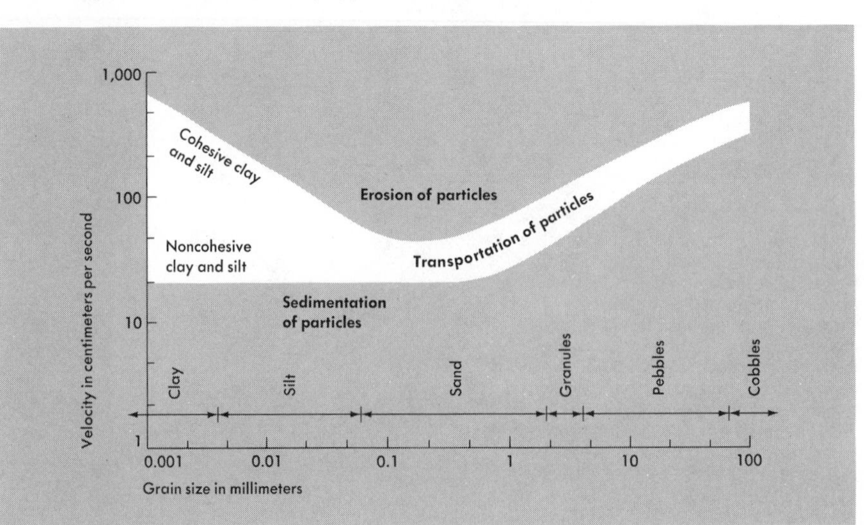

Because of this important relationship between sediment grain size and water velocity, several measures of sediment texture have been developed to differentiate deposits of one sedimentary environment from another. One such measure, called the C–M ratio, compares the coarsest fraction of a sediment with its median grain size. The coarsest fraction is the diameter of grains in the coarsest one percent, by weight, of the total sediment sample; the median grain size is the diameter of grains in the fiftieth percentile of the sample. A number of samples from specific depositional environments define characteristic fields when plotted logarithmically (Fig. 2-4). Thus, a C–M ratio of a sedimentary rock can often be assigned to a particular environment when compared to C–M ratios from known environments.

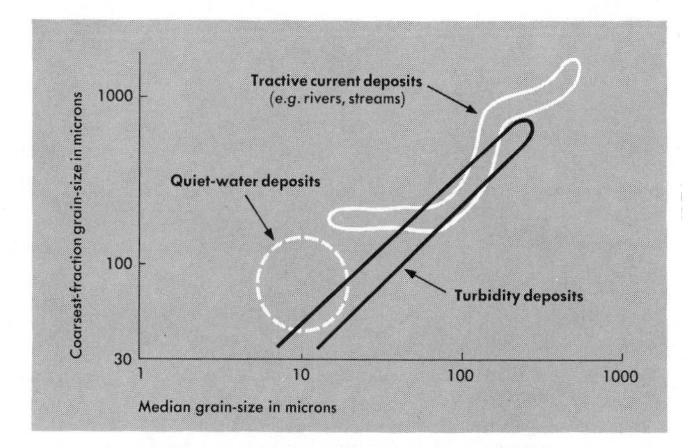

FIG. 2-4 Varying textures of selected sedimentary deposits. Although there is some overlap of the fields, each tends to characterize a certain kind of depositional environment. The fields are defined by many samples from each environment. Coordinates for an individual sample are obtained from a size analysis of that sample.

Another method for discriminating between sedimentary environments is to plot the cumulative frequency of various grain sizes from a given environment. The resulting S-shaped curve reflects the effects of the environment—especially water velocity—on the texture of the accumulating sediments (Fig. 2-5). The lower tail of the curve is the population of grains moved, carpet-like, along the sea floor, river-channel, beach, and so on. The rising part of the curve is the population of grains that bounces and hops along the bottom, part way between being bed load and being suspended load. The upper tail of the curve includes the grains carried in suspension. As with C–M ratios, comparison of an unknown sample with cumulative frequency plots of known samples may help identify the environment in which the unknown sample was deposited.

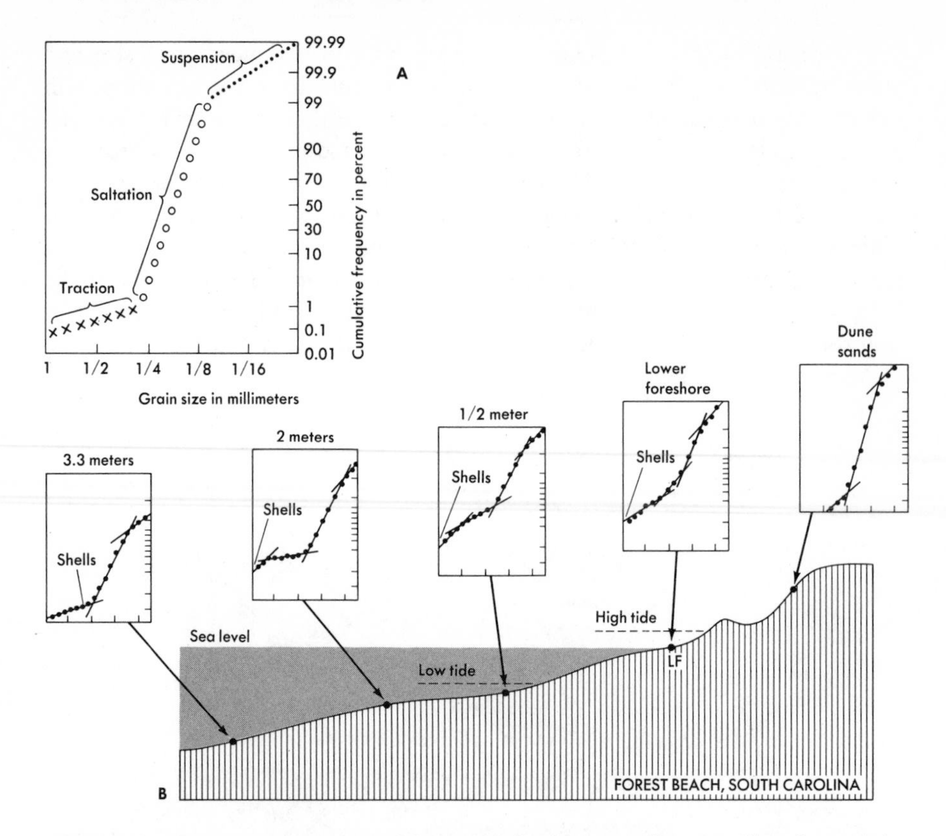

FIG. 2-5 Sediment textures measured by cumulative frequency diagrams. Plot of cumulative frequency, expressed in percent, of various grain sizes in a sediment sample yields an S-shaped figure whose lower tail includes grains carried by traction, with the rising portion including grains in saltation—bouncing and hopping along the sediment surface—and the upper tail containing the grains in suspension (upper left). The figure shows five sediment samples from a South Carolina beach, going from more than 3 meters below low tide up onto dune sands above high tide. (After G. S. Visher, 1969.)

Both these grain size measures depend, of course, upon being able to break down a rock into its individual, discrete grains. Unfortunately, many ancient sedimentary rocks are too lithified, and these methods cannot always be easily applied. However, thin-section grain-size measurements, although more time consuming, can provide good estimates of the grain sizes of lithified sedimentary rocks.

Transportation of sediments, especially by water, influences them in other ways besides determining grain size. The process of transportation causes individual grains to be rounded; finer-grained, clay-size particles to be winnowed; and the grains to be better sorted. Consequently, sediment from a freshly weathered rock may start with angular, poorly sorted grains that include clay-sized particles, but will become well-rounded and well-sorted sand after considerable water transportation (Fig. 2-6).

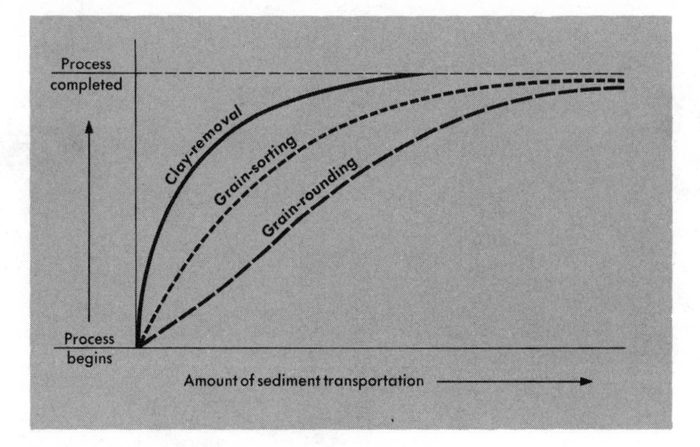

FIG. 2-6 Processes of clay removal, grain-sorting, and grain-rounding. The role of each process varies with the amount of sediment transportation.

PRIMARY STRUCTURES

We've seen how the composition and size of sedimentary grains reflect differences among various depositional environments. We want now to consider how the internal arrangement of these grains within sediments and sedimentary rocks record some of the physical and biological attributes of the environments where they are deposited. These internal grain arrangements are called *primary structures*, because they form during or shortly after the sediments accumulate. Some primary structures, such as ripple marks or animal tracks and trails, form during sediment accumulation. Others, like mud cracks or animal burrows, develop soon after sediment deposition. In either case, however, the primary structures are usually characteristic of the depositional environment, and therefore often provide internal evidence about the nature of that environment. Of course, primary structures cannot be transported from one environment to another and, therefore, they are valuable environmental indicators. Some common primary structures of inorganic origin are described here, together with their environmental significance. In the section after this we will discuss how organisms by their activity not only generate primary structures, but also contribute sediment themselves and thereby help build sedimentary rocks.

Cross-Stratification

An internal layering of sedimentary grains that is inclined to the principal surface of deposition is called *cross-bedding* or *cross-stratification*; see Fig. 2-7(A).

Because the inclination of the cross-strata, which may be up to 30 degrees, points in the direction of local current movement, it is possible to determine

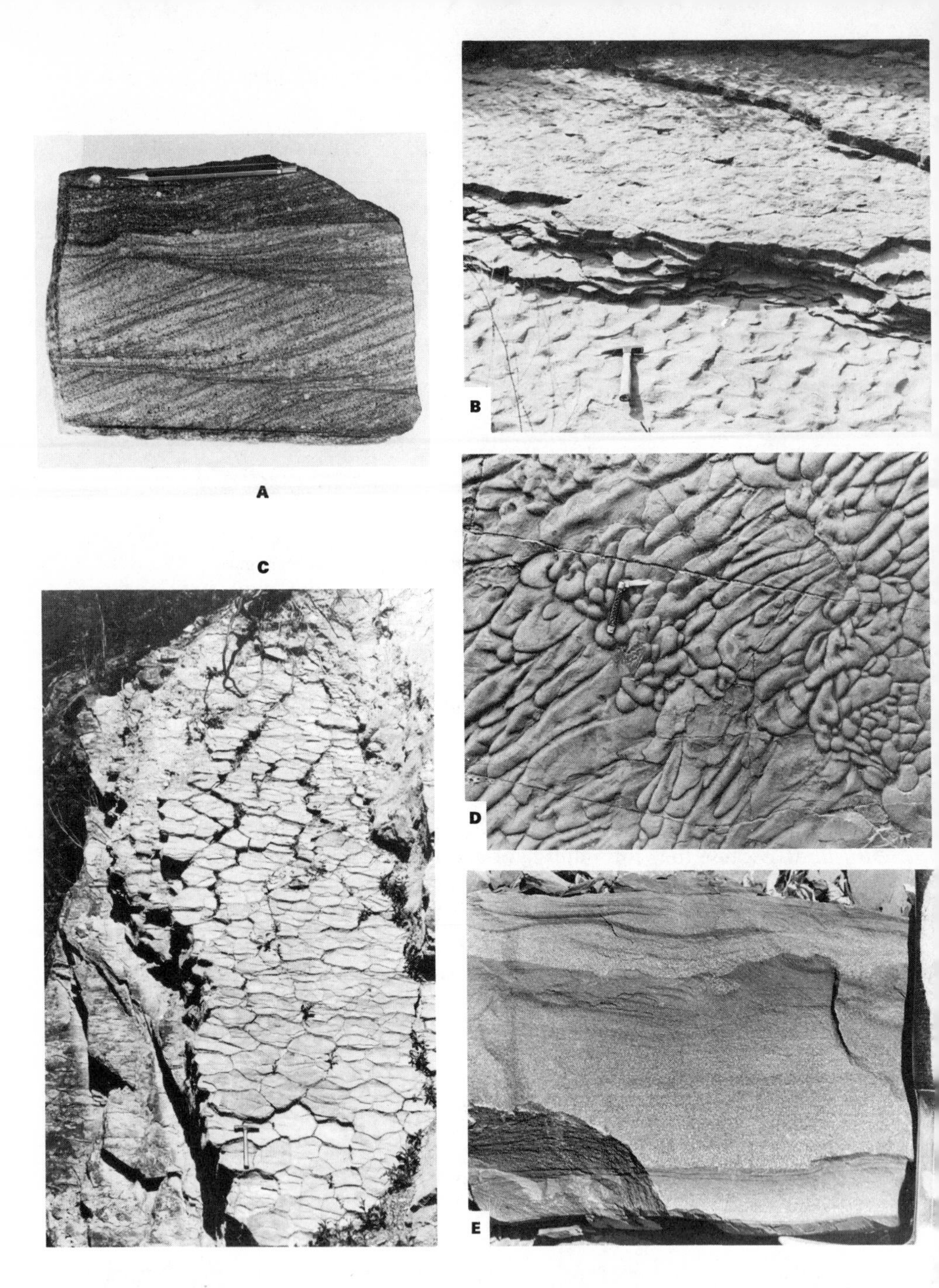

FIG. 2-7 Some common inorganic primary structures. (A) Three sets of cross-stratification in a Mississippian sandstone where current moved from right to left. (B) Asymmetrical ripple marks in a Pennsylvanian sandstone where current moved from left to right. (C) Mud cracks in a Silurian limestone. (D) Sole markings on the bottom of an Ordovician turbidite bed where current flowed from lower left to upper right. (E) Three cycles of graded bedding in an Ordovician shale. (Courtesy F. J. Pettijohn and Paul Potter, Springer-Verlag, New York, Inc.)

from an analysis of cross-stratification within a sedimentary rock unit not only the general current direction but also the direction of the sediment's source (refer back to Fig. 1-2). The geometry of cross-stratification within a sedimentary rock may vary a good deal, because wind or water currents do not always flow in exactly the same direction. But if a number of observations are made on cross-stratification, average regional directions of current flow can usually be established.

It is often possible to do more than this, however. There is a great deal of variety in the kinds of cross-stratification, depending on strength and regularity of current flow, as well as size and volume of sediments being transported. Detailed analysis of cross-stratification is thus an important part of defining and recognizing ancient sedimentary environments.

Ripple Marks

A surface of loose sediments may develop an undulating or rippled appearance as air or water currents move across it (Fig. 2-7(B)). Where the current is moving uniformly from one direction to another, the ripple marks will be asymmetrical, with their steeper sides facing downstream (or downwind), while oscillating currents will form symmetrical ripples. Like cross-stratification, asymmetrical ripple marks can be used to infer former current directions. In fact, cross-stratification is simply the cross-sectional representation of the surface rippling of a sedimentary bed. Both primary structures—one seen on the surface of sedimentation, the other at right angles to it—are generated by the same physical process, namely a moving current of water or air. As the force of the water or wind flow increases, sedimentary particles begin to move in the flow. With increasing flow, larger and larger particles move downcurrent or downwind. The surface of the sediments across which the water or wind flows becomes rippled owing to the movement of the particles (Fig. 2-8). If the flow is strong enough however, all the surface particles move together, and the sediment surface appears smooth. Of course, the surface is also smooth or flat when there is no water or wind current, or if the current is not strong enough to move any sediment.

Mud Cracks

Fine-grained, water-laid sediments that are later exposed to the air will usually shrink and crack as they dry out. These desiccation cracks form irregularly shaped polygons whose size is proportional to the thickness of the layer being dehydrated and to the drying time (Fig. 2-7(C)). Although some muds form shrinkage cracks under water ("syneresis cracks"), they do not have the connected, polygonal pattern of air-dried, mud-cracked sediments. Mud cracks, therefore, provide evidence of periodic exposure to air and, if combined with other relevant evidence, may indicate periods of temporary or long-continued terrestrial conditions.

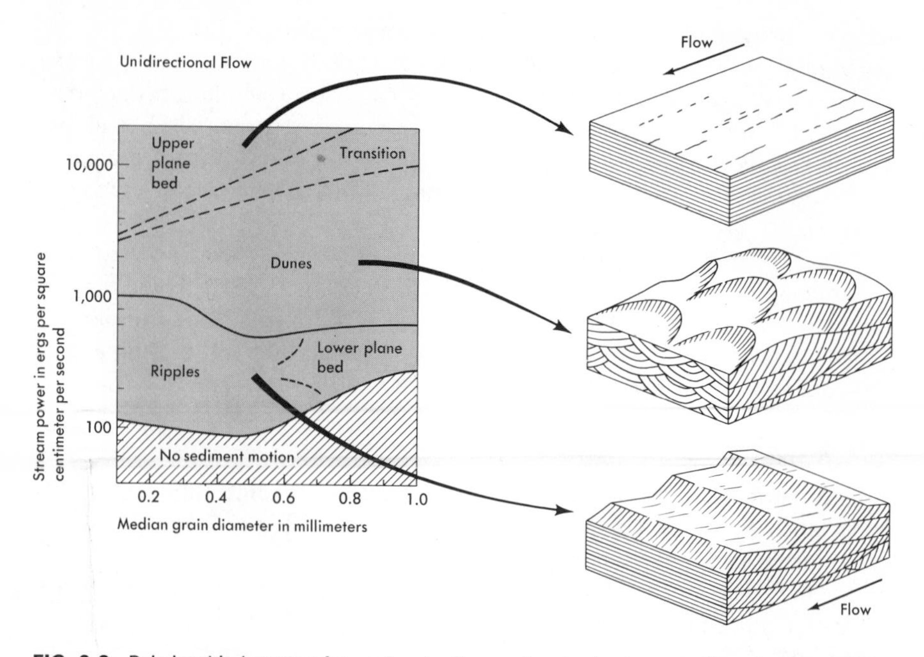

FIG. 2-8 Relationship between force of water flow and grain size in controlling the shape of the sediment surface. At low flow, there is no sediment motion; with increasing flow, the sediment surface becomes rippled and then dune-shaped; at high flow, all the sediment moves along the surface, leaving a plane, or flat, sediment surface. Not shown here are additional sediment structures that will form along the sediment surface at still higher flows. (After J. C. Ingle, Jr., 1975.)

Sole Markings

Deposition by turbidity currents may be preceded by local erosion of the sea floor as the debris-laden current moves across it. The scour marks that are formed by the turbidity current have a variety of characteristic forms, some of which indicate the direction from which the current came. The scour marks are actually preserved as casts formed by the sediment that is deposited by the turbidity current after its eroding front has passed by. The casts of these scour marks thus appear on the sole, or bottom, of the turbidity-laid sedimentary layer, hence the term "sole markings" (Fig. 2-7(D)). Other sole markings include local casts where the contact between sand and mud layers are deformed during compaction.

Graded Bedding

The sediment deposited by a turbidity current is usually laid down in such a way that the coarse grains are dropped first, followed by the settling out of the finer grains. This change in texture occurs because as the turbidity current's velocity decreases, the coarsest pebbles and sand grains will be deposited initially. This coarse layer will then be slowly buried by finer-grained sand, silt, and clay as they settle out of suspension from the overlying turbid water. This regular variation in grain size from the base of a sedimentation unit to the top is called *graded bedding* (Fig. 2-7(E)). Although graded bedding can also be found in other sedimentary environments, it is quite typical of rock sequences laid by turbidity currents, and combined with other criteria, such as sole markings, can be useful in defining turbidity-current environments.

Figure 2-9 illustrates the vertical and lateral variation seen in a turbidite bed from its base to its top, from near its source (proximal) to some distance away from the source (distal). Graded bedding is best seen in unit A, which contains the coarsest sediment, usually medium to coarse sand. Unit B, composed of finer sand, shows parallel laminations and reflects deposition in a waning, but still strong, flow of water. (This is the "upper plane bed" of Fig. 2-8.) As the turbidity current further slackens with time and distance, unit C is deposited as a current-rippled, cross-stratified sand. (Refer again to Fig. 2-8.) Unit E records deposition of fine-grained sediments from suspension in the overlying water column following turbidite deposition. Unit D, then, is transitional from turbidite to pelagic deposition. Notice the correspondence between temporal and lateral changes in sediments as an important environmental parameter—

FIG. 2-9 Vertical and lateral changes in grain size and primary structures in a turbidite bed. The actual thickness of the bed varies, on the average, from 10 centimeters to one meter near the source (proximal), and one to 10 centimeters some distance downcurrent (distal). The sediment grains in units A, B, and C are moved by the turbidity current itself along the depositional surface. The triangular plot in the upper right shows how the average grain size in each of these units reflects the declining power of the turbidity flow. The sediments in units D and E are pelagic; that is, they settle out of suspension from the overlying water column. See text for further discussion. (After J. R. L. Allen, 1970.)

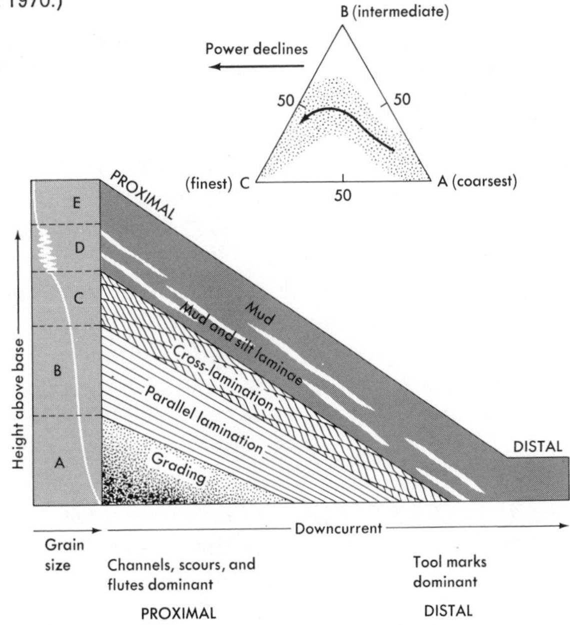

water energy—varies. We will discuss other examples like this again, where a vertical sequence of rocks records through time the same sort of environmental change one might see geographically at one moment in time.

ORGANIC INFLUENCES ON SEDIMENTS

Animals and plants can influence the composition and structure of sediments in several different ways. They contribute sedimentary grains by their skeletal debris. They also form organic primary structures by burrowing into and through sediments and, in the process, may destroy pre-existing inorganic primary structures like cross-stratification or horizontal bedding planes. Finally, organisms can build structures of various sizes like coastal marshes and coral reefs that can significantly modify local or regional patterns of sedimentation. Although we usually think of the physical environment controlling the behavior of organisms, it is just as true that animals and plants can affect the physical environment.

Organic Contribution to Sediments

Organisms contribute directly to sediments by producing, in the case of animals, a variety of internal and external skeletal materials such as bone, teeth, shells, and in the case of plants, woody tissue, all of which become sedimentary grains after the death of the organisms. Of these, however, only the calcareous- and siliceous-secreting protistans (such as diatoms, radiolarians, and foraminifers), and the calcareous algae and invertebrates (including corals, brachiopods, bryozoans, molluscs, and echinoderms) have any, real quantitative significance in the sedimentary record. For example, limestones, variously estimated to make up 10 to 20 percent of sedimentary rocks, are virtually all biologic in origin. Coal, too, which provides much of the world's energy is a biologic sediment. Even today, large parts of the Earth's sea floor are covered by a thin veneer, up to tens of meters thick, of calcareous sands and silts derived over the last several million years from the skeletons of microscopic, floating single-celled organisms.

Skeletal materials are weathered, transported, and deposited much the same as inorganically formed rocks and minerals are. Most skeletal structures are secreted in an organic matrix that decomposes after the death of its owner. The organic matrix is attacked by microorganisms and is oxidized by oxygen in the atmosphere or in water. Consequently, the individual crystalline units, which compose the skeleton and are imbedded in this matrix, are liberated and shed into the sediment. For example, a large clam shell which begins as a large, cobble-sized sedimentary grain will, as the binding matrix of the shell is removed, eventually break down into many thousands of tiny calcite prisms and aragonite needles just a few microns long. Depending on the degree of skeletal decomposi-

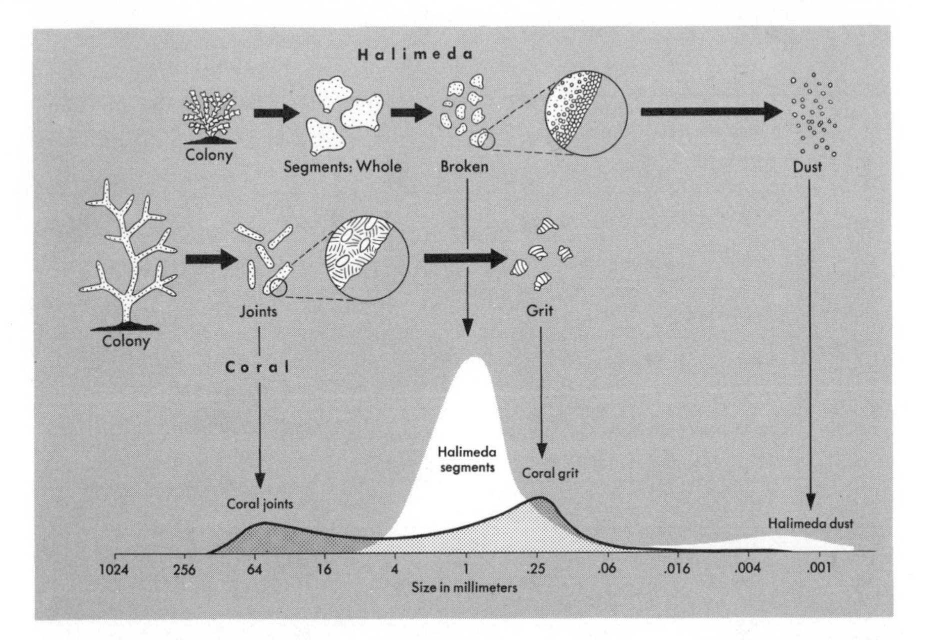

FIG. 2-10. How a green calcareous alga (*Halimeda*) and a stony coral (*Acropora*) contribute skeletal debris to sediments. Note the wide variation in size from a few centimeters to several micrometers, depending on the degree of disarticulation and disintegration of the original skeleton. These two organisms thus can produce calcareous sediments of various textures from coral gravel to lime mud. (After R. Folk and R. Robles, 1964.)

tion, shelly sediments will thus have textures that reflect the original internal architecture of the shell materials contributed by the local organisms (Fig. 2-10).

After deposition, the organically produced crystalline remains are susceptible to solution, particularly some of the more unstable mineral types, such as aragonite and opal, and provide a reservoir of calcium carbonate and silica for subsequent sediment cementation. Differing solubility of various kinds of skeletal debris can further bias any estimate about the original composition of the local flora and fauna based on fossil remains within a sedimentary rock. (The initial bias introduced, of course, is that of the nonpreservation of the many soft-bodied organisms that secrete no mineral material whatsoever.)

In short, then, through their secretion of various kinds of skeletal materials, organisms can contribute directly and significantly to the ultimate composition and texture of a sediment. And because organisms in turn are limited in their distribution and abundance by the local environment (as will be discussed in the next chapter), the influence that the depositional environment exerts on the character of the accumulating sediments is demonstrated once again. Table 2-2 shows the distribution of skeletal minerals in the principal groups of animals and plants.

Table 2-2 Distribution of Skeletal Minerals in the Principal Phyla of Organisms

	Phyla		Calcium carbonate $CaCO_3$		Opaline Silica $SiO_2 \cdot nH_2O$	Calcium Phosphate $Ca_5(PO_4)_3OH$
			Calcite	Aragonite		
Plants	Schizomycophyta (Bacteria) Pyrrophyta (Dinoflagellates)					
	Cyanophyta (Blue-green algae)		Frequent			
	Chlorophyta (Green algae)			Frequent		
	Charophyta (Stone worts)		Frequent			
	Phaeophyta (Brown algae)					
	Rhodophyta (Red algae)		Common			
	Chrysophyta	Diatoms			Common	
		Coccolithophorids	Common			
	Mycophyta (Fungi) Bryophyta (Mosses) Tracheophyta (Vascular plants)					
Animals	Sarcodina	Radiolarians			Common	
		Foraminiferans	Common			
	Porifera (Sponges)		Frequent		Common	
	Coelenterata (Corals)		Common	Common		
	Bryozoa (Bryozoans)*		Frequent	Common		
	Brachiopoda (Brachiopods)		Common			Frequent
	Mollusca	Snails*	Frequent	Common		
		Clams*	Common	Common		
		Cephalopods		Common		
	Annelida (Segmented worms)*		Frequent	Frequent		
	Arthropoda	Trilobites Crustaceans	Common			Common
		Arachnids Insects				
	Echinodermata (Echinoderms)		Common			
	Chordata	Acorn worms Tunicates, lancelets				
		Vertebrates				Common

Rare or absent

Frequent

Common

* Mixed calcite-aragonite skeletons often occur in these groups.

(After A. L. McAlester, 1977.)

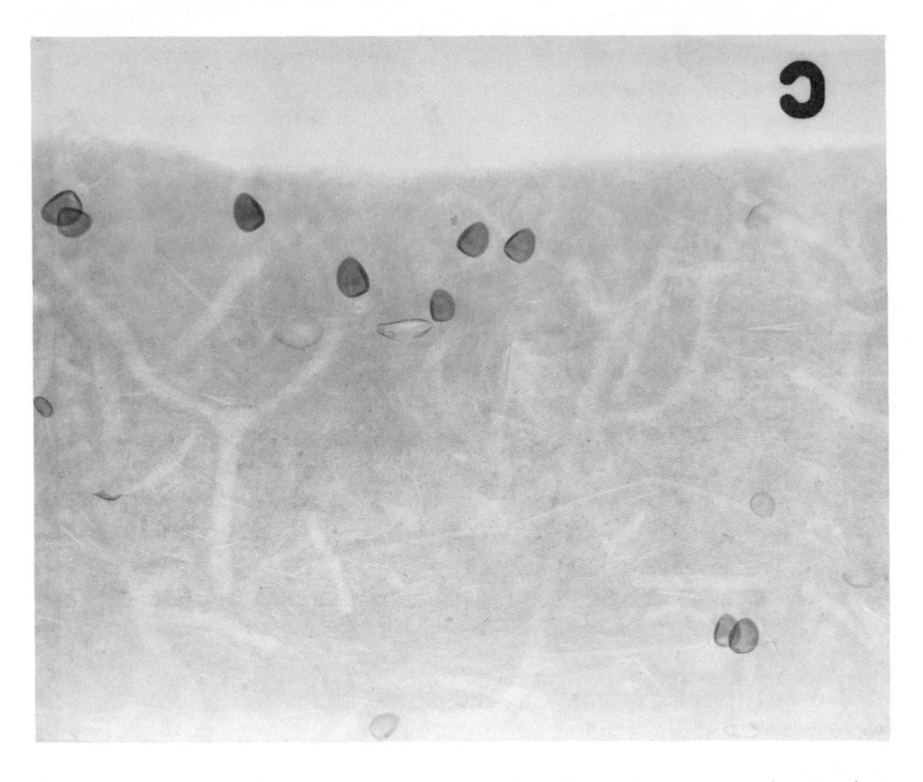

FIG. 2-11 X ray of marine sediments off southern Massachusetts in water several meters deep. Burrows made by a variety of soft-bodied and shell-bearing invertebrates including crustaceans, polychaete worms, and bivalves.

Biogenic Structures

Organisms contribute in other, less direct ways to sediments. Many marine organisms, especially worms, arthropods, and molluscs, burrow into sediments for shelter and food (Fig. 2-11). By so doing they disrupt bedding, obliterate primary structures, and increase sea water circulation within the sediments. Organisms that feed on sediment for its included organic matter may aggregate the sediment to form pellets as it passes through the digestive tract (Fig. 2-12). In environments where sediment is reworked by organisms faster than it accumulates—because of either slow sedimentation rates or large populations of organisms—the resulting sediments may be so extensively modified that they retain none of the original, inorganic primary structures (Fig. 2–13).

Besides modifying the character of sediments already deposited, organisms are also capable of altering patterns of sedimentation. For instance, thin mats composed of the intertwined filaments of various kinds of blue-green and green algae can exert local, small-scale influence. These algal mats, which occur in very shallow water, both fresh and marine, as well as on moist land surfaces,

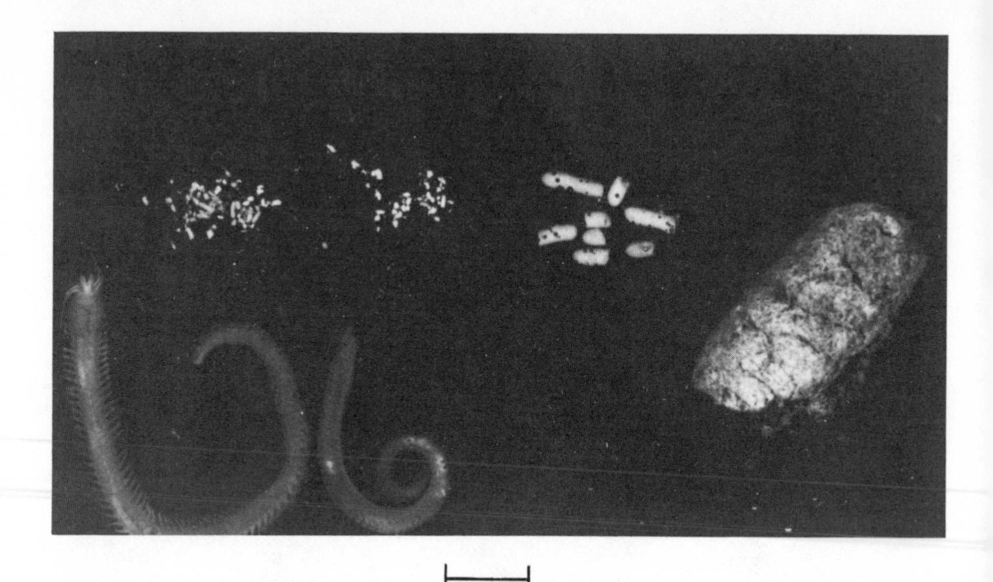

FIG. 2-12 Representative fecal pellets formed by various marine invertebrates. The two examples at the upper left are formed by worms, the pellets in the center are from an intertidal mollusc (chiton), while the large pellet at the right is from a crustacean. (Courtesy R. N. Ginsburg, 1957.)

form tough, leathery coverings on the sediment surface and thus inhibit or prevent the transport of grains along the surface. Hence, the mats act to keep sediments stable and free from erosion after their initial deposition.

Algal mats are also capable of trapping and binding sediment grains on their gelatinous surface. As the mats are covered by newly deposited sediments, the algal filaments grow upward and form yet another mat. Successive periods of matting and deposition can result in a well-laminated sediment (Fig. 2-14). Furthermore, the geometry of these laminations is related to the frequency and strength of local water movement.

Some of the oldest sedimentary rocks known contain laminated structures that are virtually identical with recent algal laminated sediments. These rock structures, termed *stromatolites*, are believed to be algal in origin, the layering or laminations being related to the successive development of algal mats, although the algal tissue itself has not been preserved (Fig. 2-14). The presence of stromatolites in many Precambrian rocks, some of which are dated at 2 billion years or more, indicates the great antiquity of life—even if only "primitive" life. Stromatolites are also useful environmental indicators because rocks containing them were presumably deposited in very shallow water, if we can assume that algal mats occurred in the same environments in the past as they do today. In many cases this assumption appears justified, because algal stromatolites occur in close association with mud cracks and intraformational conglomerates, indicating subaerial exposure, desiccation, and local erosion of the accumulating sediments.

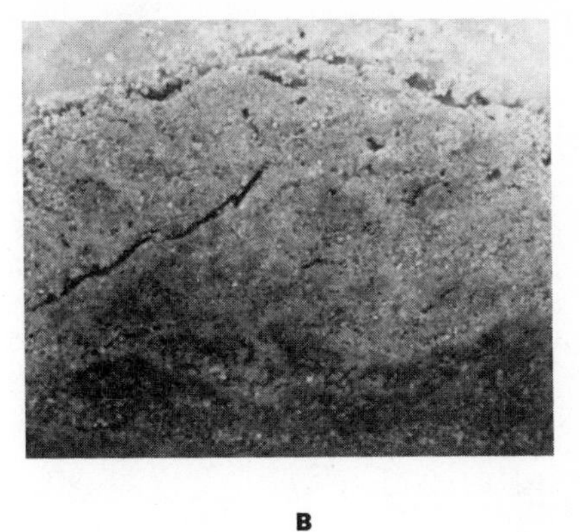

A

B

FIG. 2-13 (A) Artificially laminated sediment with alternating layers of fine (dark) and coarse (light) sand in an aquarium. (B) Five marine worms were placed within the aquarium and after one month the four-and-one-half-inch bed of sediment has been stirred up and the bedding has been obliterated. (Courtesy R. N. Ginsburg, S.E.P.M. Special Publication No. 5.) (C) Natural view of a marine echinoid burrowing through a rippled, calcareous sand. (Courtesy Norman D. Newell, American Museum of Natural History, S.E.P.M. Special Publication No. 5.)

C

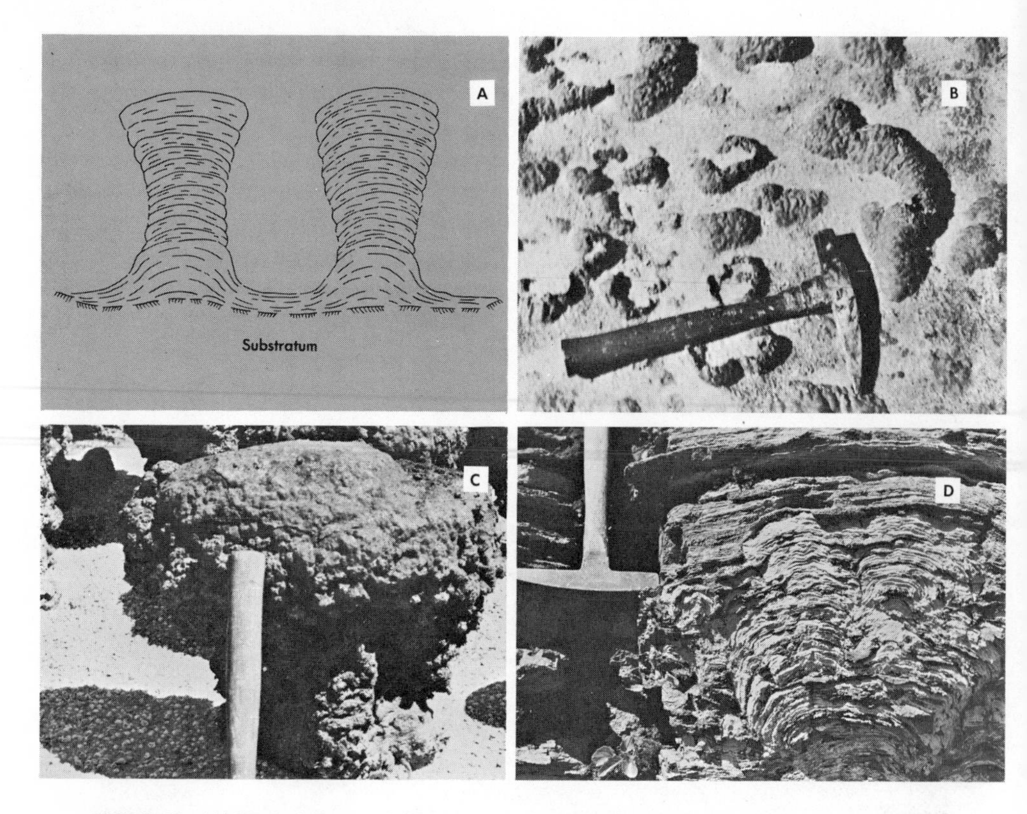

FIG. 2-14 (A) Vertical sketch of algal laminated sediment, algal mats localized as small mounds which grow upward as successive layers of sediment are trapped by successive generations of mat formation. (B) A number of small algal mounds on the floor of a saline lake in western Australia. During part of year the lake floor is covered with water and the algal mats are rejuvenated; hammer gives scale. (Courtesy Brian W. Logan.) (C) Large, single algal mound from the intertidal zone of Shark Bay, Australia; coarse shell debris surrounds the mound. (Courtesy Brian W. Logan.) (D) Cross section of a fossil algal mound (or stromatolite) from Precambrian rocks, about 1 billion years old, of Montana. (Courtesy Richard Rezak, U.S. Geological Survey.) (E) Series of events, from left to right, in the development of an algal stromatolite. (From H. Hofmann, 1969.)

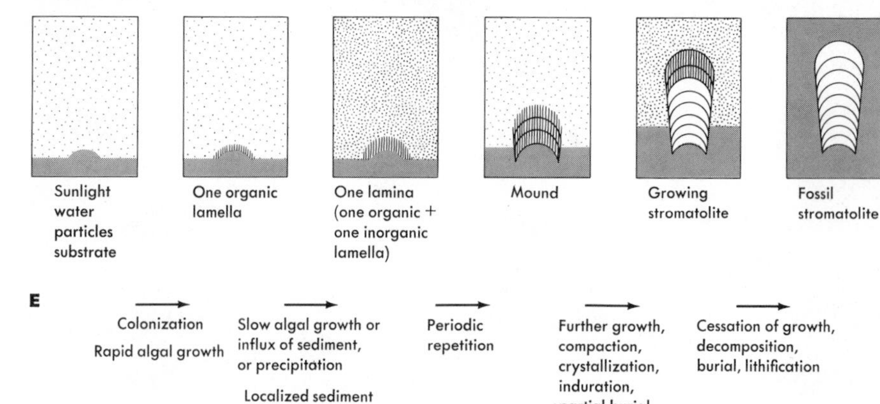

Other, and geologically more dramatic, examples of the organic modification of local patterns of sedimentation are provided by reefs. Organic reefs developed in those parts of ancient seas that were clear, shallow, warm, well lit, and agitated, so that a profusion of shelly marine invertebrates flourished. Some of these organisms—such as stony corals, calcareous sponges, and certain rooted brachiopods—acted as *frame-builders* of these reefs. Other organisms—such as calcareous algae and hydrocoralline coelenterates—because of their encrusting mode of growth acted as the *cement*, binding the framework of the reef together into a rigid, wave-resistant structure. Other marine invertebrates—such as molluscs and echinoderms—provided the *detrital fill* for the growing reef mass (Fig. 2-15(A)).

Reefs can perpetuate the favorable conditions for their initial formation by growing as fast as the sea floor subsides, thereby maintaining the survival of the reef community. With time the reefs become large masses of calcareous rock surrounded by nonreef sediments. In some cases the reefs will form long, linear bodies that build up to mean sea level. This type will have a quieter water lagoon behind the reef-wall proper, and a turbulent fore-reef environment. Thus, what began as an essentially uniform sedimentary marine environment is gradually transformed by reef development into several highly differentiated marine environments: *back-reef lagoon*, *reef barrier*, and *fore-reef slope* (Fig. 2-15(B)).

Significance of Organic Influence on Sediments

Having established that organisms contribute grains to sediments as well as disrupt their internal structures, we may reasonably ask how geologically significant such activity is. That is, given the rate at which these various organic processes occur, how great will the effects be in the geological record?

Sedimentary rocks constitute about 75 percent areally and 5 percent volumetrically of rocks within the Earth's crust. As just noted, limestones—which are almost entirely biologic in origin—are thought to account for anywhere from 10 to 20 percent of the total. Hence, of the estimated 400 million cubic kilometers of sedimentary rock, 40 to 80 million are the result of calcareous-secreting and -precipitating organisms. Although this is a startling figure in absolute terms, given the great lapse of geologic time, slow depositional rates can easily account for such a great mass of biogenic calcareous sediments. For example, the present-day Bahama Islands in the Caribbean are underlain by a section of shallow-water, shelly sediments that are at least 4,500 meters thick and that go back to the late Cretaceous Period. During the last 90 million years, therefore, some 4,500 meters of limestone have been deposited, giving an average accumulation rate in this area of only about 5 centimeters per 1,000 years. Such a rate surely must be minimal, however, because sedimentation has not been continuous owing to times of lowered sea level—as during the Pleistocene glaciations—when the Bahamas stood above water and were subjected to erosion. Present sedimentation rates as measured since the last inundation of the shallow Bahama platform indicate that the true rate may be as much as ten times as great.

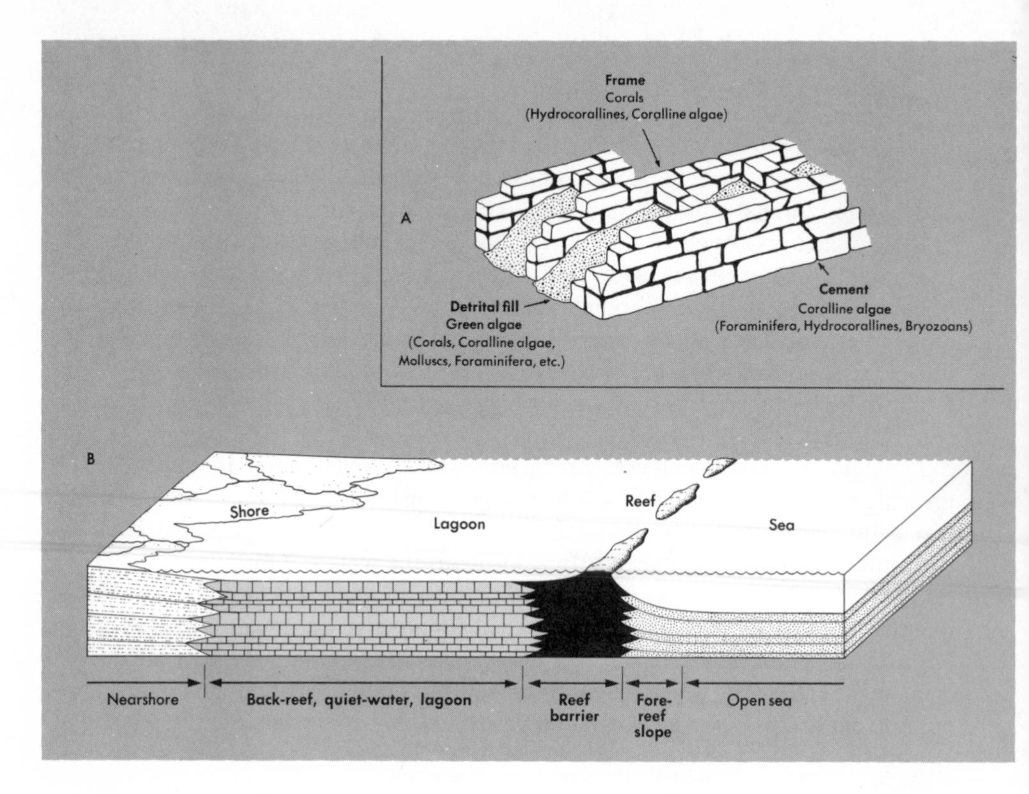

FIG. 2-15 (A) Schematic drawing of the role of various calcareous organisms in reef construction. (After R. N. Ginsburg and H. A. Lowenstam, 1958.) (B) Ancient reefs often had different kinds of organisms but they performed much the same structural function. As a reef develops, local patterns of sedimentation are altered; major subenvironments within a reef complex are the reef barrier and the environments in front of (more turbulent) and behind (less turbulent) the barrier. (C) Oblique aerial view of barrier reef, Belize, Central America. Deep water of the Caribbean to right; reef barrier, center; and shallow lagoon to left. (From R. K. Matthews, 1974.)

Estimates can also be made for accumulation rates of the minute calcareous shells of floating protistans that rain down on the ocean floors. For example, a core raised from deep-water sediments from an area south of the island of Hispaniola measures more than 900 centimeters, representing sediment accumulation during about 400,000 years. An average sedimentation rate for this Caribbean core is thus calculated at almost 2.5 centimeters per 1,000 years.

How do these shallow and deep-water rates of carbonate sedimentation compare with areas where no carbonate is deposited? In the Gulf Coast sedimentary basin, sedimentation rates are estimated at 20 centimeters per 1,000 years as averaged over the Cenozoic Era. Once again, this is probably too low a figure because this area has, at times, been subject to erosion and so has not continuously received sediments. Depositional rates for red clays in the deep ocean basins are quite low, ranging from one-twentieth to one-half centimeter per 1,000 years. Sediments of organic origin are indeed being deposited at geologically significant rates that are comparable to those for inorganic marine deposits.

In several instances marine biologists and geologists have measured the rate at which sediment is *re*deposited by various marine organisms. Often these rates are of the same order of magnitude as primary rates of inorganic and organic deposition. For example, many of the reef-dwelling fish of Bermuda (parrotfish, triggerfish, puffers, and the like) browse on the calcareous sand of the reef for its included food and for use as a milling agent in grinding the algae they have eaten. These fish annually redeposit between two and three metric tons of calcareous material per hectare (about one ton per acre) on a typical Bermuda reef. These data convert to a sedimentation rate of 10 to 15 centimeters per 1,000 years. Similar observations of some Pacific reefs yielded redeposition rates by fish of 20 to 30 centimeters per 1,000 years.

Burrowing marine invertebrates can also redistribute and redeposit materials after their initial deposition. Various studies—for instance, of worms, holothurians (sea cucumbers), and clams—indicate that these organisms are capable of reworking sediment just as fast as it is deposited by inorganic agents.

It seems clear, then, that organisms do make truly significant contributions to the sedimentary rock record through the formation of skeletal sedimentary grains as well as through the reworking and redeposition of sediment. In fact, when we once realize how effective organisms can be as rock-builders and sediment-modifiers, we may ask why *all* sedimentary rocks don't show these organic influences.

Some sedimentary rocks, such as evaporites, glacial deposits, and dune sandstones, will, of course, lack traces of any organic activity because they are formed in environments where organisms are virtually excluded. Other depositional environments may contain rock-building and sediment-modifying organisms, but the rate of organic activities may be far less than the rate of sediment influx and accumulation, so that the sediments are buried more rapidly than they can be effectively reworked by organisms. The production of calcareous sediments—either as skeletal sands, shell beds, or reefs—requires that only limited amounts of noncarbonate land sediments come into the sedimentary basin so that the organically produced sediments are not diluted. Hence, preservation of either inorganically produced primary structures (cross-stratification, ripple marks, graded bedding, and so on) or organically produced structures (burrows, pellets, shell beds) depends on the relative rates at which the inorganic and organic processes are working. For example, in turbulent shoal areas along the margins of the Great Bahama Bank, the calcareous sediments retain their primary current stratification because of the constant reworking by tidal flow across the shoals and because of the relatively few bottom-dwelling burrowing organisms that might disturb the sediments. On the other hand, the calcareous sediments from the interior lagoon of the Great Bahama Bank lack any primary stratification and, in the absence of any regular, strong bottom currents that could restratify the sediments, are thoroughly stirred up by the abundant burrowing fauna found there (Fig. 2-16).

FIG. 2-16 Box cores from the uppermost 30 centimeters of calcareous sands from the Great Bahama Bank. The core on the left shows excellent primary current stratification as developed on the foreshore of a beach. The core on the right shows no primary stratification and is heavily burrowed by marine animals living in these lagoon sediments. (Courtesy J. Imbrie and H. Buchanan, S.E.P.M. Special Publication No. 12.)

DEPOSITIONAL SYSTEMS AND THE STRATIGRAPHIC RECORD

At the moment of deposition, sediments have areal dimensions that are, of course, closely related to the nature of the depositing medium. Stream and beach deposits are linear, running parallel to the direction of stream flow or surf action. Deep-sea muds and lagoon sands will be blanket-like and usually widespread. Reefs and evaporate deposits are often quite discontinuous or patchy in their occurrence, and so their distribution is localized and irregular. With time, however, this two-dimensional geometry acquires a third dimension, because as sedimentation continues, a significant thickness of sediment accumulates. Thus, the nature of individual rock strata informs us about the particular depositional environment that formed it, whereas the vertical sequence of strata tells us what temporal changes there might have been in the depositional environment. As noted in Chapter 1, paleoecologists are as much interested in these changes in environments over time as they are in changes in environment over a geographic area at a specific moment. In this section, therefore, we want to consider the paleoecological interpretation of depositional sequences seen in the stratigraphic record, both in terms of the nature of individual strata and in terms of their variation through time.

The Facies Concept

Within a basin of sedimentation there are, usually, a number of different, local depositional environments. These local environments reflect variations in physical, chemical, and biological conditions within the basin as well as their distance and direction from any depositional agent that may be entering the basin—such as a river with its associated delta. Hence, at any one time the sediments being deposited throughout the basin will have different characteristics and overall aspects that are correlated to the local depositional environments. Such lateral variations within a sedimentary basin are termed *sedimentary facies*. The depositional sites of these individual facies may also shift their position through time so that each facies will have its own three-dimensional configuration in the total stratigraphic sequence deposited within the basin as a whole (Fig. 2-17).

Earlier in this chapter we indicated how sediments will have inorganic and organic attributes of texture, composition, internal structures, fossils, and so on. Correspondingly, sedimentary facies may be characterized from their inorganic, lithologic qualities—*lithofacies*—as well as from their biologic qualities—*biofacies*. In either case, lithofacies and biofacies are a direct manifestation of the local depositional environment within a sedimentary basin. In order to define and interpret the origin and history of a given stratigraphic sequence, it is useful to identify first the various sedimentary facies contained therein. Instead of speaking of "20 meters of upper Cretaceous sandstone in this county and 50 meters of shale in the next" we may say "nearshore quartz sands here, passing

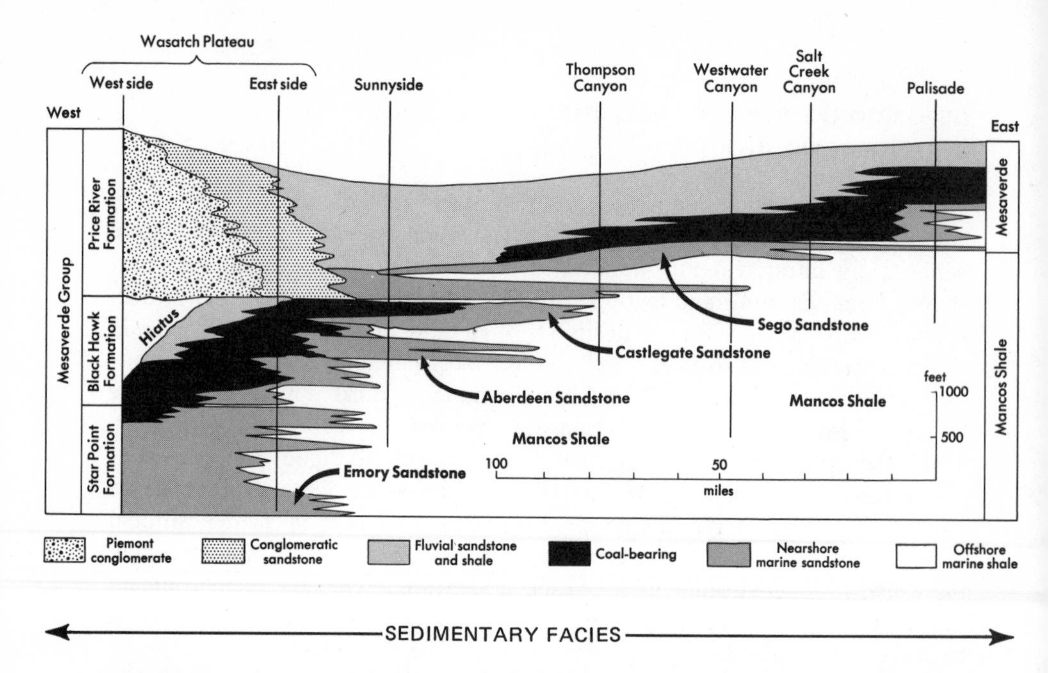

FIG. 2-17 Cross section of the Mancos Shale and Mesaverde Group of eastern Utah and western Colorado. These Upper Cretaceous deposits record nonmarine and nearshore marine sediments (Mesaverde Group) interfingering eastward with offshore marine deposits (Mancos Shale). The stratigraphic section as a whole indicates gradual withdrawal of the sea to the east as erosional debris enters the area from the west during the uplift of the ancestral Rocky Mountains. Note the complex intertonguing of different sedimentary rock types, or facies, which records constant shifting of local depositional environments with time. (After C. O. Dunbar and J. Rodgers, 1957.)

into offshore shallow marine muds over there." The recognition and interpretation of facies thus permit us to move from a purely descriptive narrative of what is seen in scattered rock outcrops (certainly a necessary first step) to a discussion of how and why certain kinds of rocks accumulated where they did, what their relationships are to each other, and what sorts of life habitats might have existed for any associated organisms.

The recognition and interpretation of lithofacies and biofacies is the crucial goal of environmental stratigraphy, which we discussed in Chapter 1. We will return to the facies concept again in Chapter 5, where we discuss in more detail the whole subject of environmental analysis. What we want to emphasize about the facies concept here, however, is the important realization that depositional environments at a given moment are not spatially uniform. On the contrary, they do vary from place to place, and this variation is recorded by the different sediments and organisms found in each specific environment. As a result of this lateral variation in environment, we must always expect a particular sedimentary rock in the geologic record to change in facies, sooner or later, as we trace it laterally from one place to another. The facies concept is counter to an earlier, now outdated, notion often referred to as the "layer-cake concept," whereby rock strata were visualized as relatively homogeneous layers that recorded uniform environments.

Walther's Law

In 1894, the German geologist Johannes Walther published an important book in which he noted that different kinds of sediments are deposited adjacent to each other in space owing to lateral variations in depositional environments. Therefore, a vertical sequence of different sedimentary rocks must record the superimposition of those same environments over time. In other words, the sequence of rocks seen at one particular place provides evidence not only that the depositional environment changed through time there, but also evidence of what the lateral variations in environment were in the general region. For example, refer back to Fig. 2-17. The Upper Cretaceous rocks exposed at, say, Salt Creek Canyon indicate a change in depositional environment from offshore marine muds to nearshore marine sands to coastal swamps to fluvial, nonmarine sands. Moreover, the vertical superposition of the rocks is such that each successive rock stratum, or facies, was the lateral equivalent of the one below and the one above. Furthermore, given the origin of these different facies, we can conclude that this interval of the stratigraphic record on the Colorado Plateau records a regional retreat of the Cretaceous sea. (Why?)

Acceptance of the validity of Walther's Law has led to the further recognition that when we look at sedimentary rocks in the field, we should pay as much attention to the superimposed vertical sequence of rock types, or facies, as to the lithologic and paleontologic character of an individual rock stratum. Although we may not always be able to interpret correctly the depositional environment of one rock layer or another, our chances of success improve greatly if we look at the whole package of strata which, altogether, records the lateral migration of a broad environmental regime through the area.

In fact, this approach to interpreting the stratigraphic record has yielded valuable results for paleoecology. There are, to date, more than a dozen kinds of vertical sequences of sedimentary strata that record broad environmental regimes, with each stratum or layer in the sequence reflecting a specific depositional environment. As you might guess, these *facies models* provide extremely useful interpretive guides to the stratigraphic record. We will discuss several of them in the following section.

Five Facies Models

We will consider five different broad environmental regimes: fluvial, deltaic, clastic tidal flat, carbonate tidal flat, and carbonate shelf. For each, we will describe the lateral variations in facies and how the facies appear in the stratigraphic record when vertically superimposed on each other. Remember, though, that these are idealized models, which are by no means always seen this perfectly in real life. But the models do represent the more important sedimentologic and paleontologic features abstracted from many specific examples studied by geologists.

Fluvial facies. One major sedimentary environment on dry land is that of the river valley. Most terrestrial erosion occurs by flowing water, and the sediments generated thereby are carried down stream in river channels, eventually reaching the sea or an inland lake. In the lower reaches of a typical river valley, the river meanders back and forth over the sediments that it has previously deposited. Over time, the river erodes and transports this sedimentary material, all the while lowering its longitudinal profile from its mouth, where it empties into the sea or lake, to its headwaters, in the surrounding upland areas.

As shown in Fig. 2-18, the sedimentary deposits in the river valley include various kinds. First, there are the coarse sands and gravels in the channel itself that are part of the river's bedload (refer back to page 16). Next, there are the sands deposited on the inside bends of the meandering channel; these are called *point bar deposits.* The river erodes sediment on the opposite, or cut, bank on the outside bend, and it migrates, or meanders, in that direction.

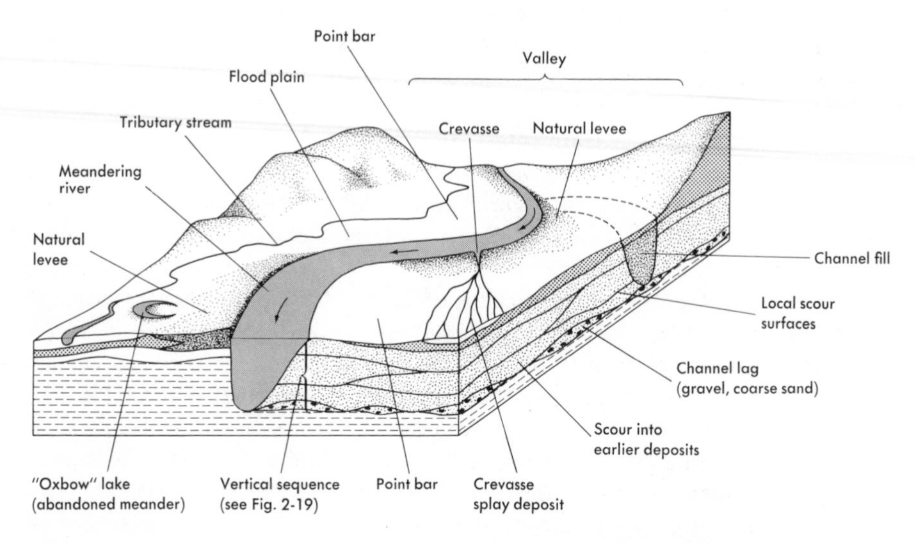

FIG. 2-18 Block diagram of a meandering river valley. The river flows back and forth across its own flood plain, leaving in its wake a vertical sequence of characteristic fluvial facies, which is shown in detail in Fig. 2-19.

At times of high water the river will occasionally rise over its channel banks and flood the adjacent valley. As the rising water leaves the channel, it loses some of its flow velocity and, consequently, deposits sand and silt along the margins of the channel, forming natural levees. The finer-grained, suspended sediment settles out of the flood waters in the valley. Sometimes flood waters breach a previously formed natural levee and create a *crevasse splay deposit* on the valley floor next to the river channel.

What we have just described is how the river channel and its sediments look at one particular short interval of time. Turning to Fig. 2-19, you will see

STRUCTURES/TEXTURES	ENVIRONMENT
Siltstone and mudstone: poorly-bedded; mudcracked, rooted; peaty horizons	Overbank flood plain
Sandstone and siltstone: ripples, root-mottling; sheet-like geometries	Levee/crevasse splay
Sandstone: fine-grained, ripple-laminated	Upper point bar
Sandstone: fine-to medium-grained; flat-bedded, laminated	
Sandstone: medium-grained, cross-bedded; common trough sets and transverse sets on slip-off face; armored mud balls	Lower point bar
Conglomerate and coarse sandstone: scour-fill	Channel lag

FIG. 2-19 Generalized vertical sequence, or fining upwards cycle, of fluvial facies in a meandering river valley. Compare with Fig. 2-18. Why do the sediments decrease in grain size upward in the sequence? Why is the sequence cyclic?

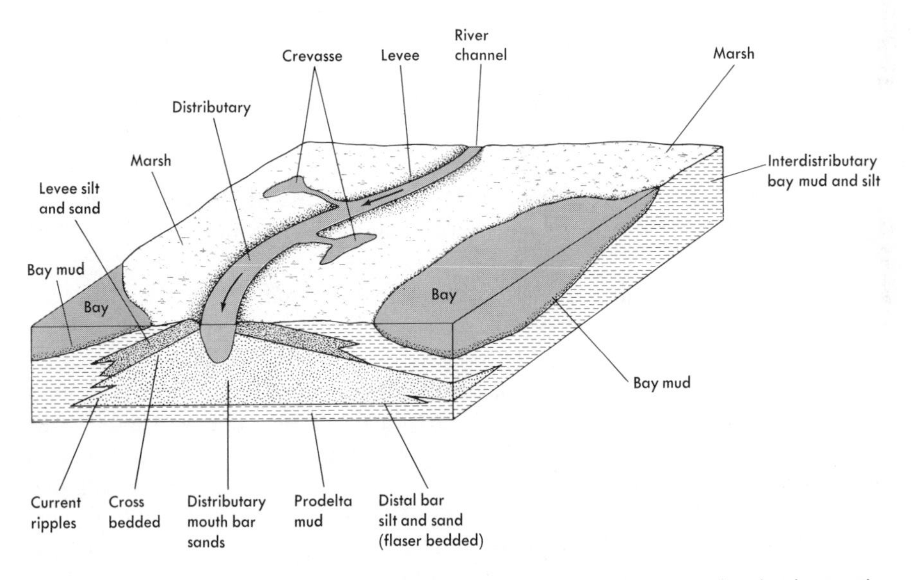

FIG. 2-20 Block diagram of a delta. Sand is deposited along the river channel and at its mouth as it enters the sea. Mud accumulates marginal to the channel and in front of the delta. Floods will create levees and crevasse splay deposits. When the deltaic sediments build up to sea level, swamps will form. Compare this figure with Fig. 2-18 which shows a somewhat similar distribution of facies further up the river channel, away from the river mouth.

the vertical sequence of sediments built up over time as the meandering river continues to migrate back and forth across the valley. Such vertical sequences of fluvial facies are commonly seen in the stratigraphic record. Note the systematic decrease in overall grain size going up in the sequence, resulting in its being termed a *fining–upwards sequence*—or cycle, because of repeated migrations of the river channel. Besides changes in sediment grain size, you will also note changes in the kinds of primary, inorganic structures, especially cross-stratification. Fossils are relatively few in such vertical sequences. When they do occur, they include isolated vertebrate bones and teeth as well as plant fragments.

Deltaic facies. When a river reaches the sea its current velocity drops rapidly and, consequently, the sediments carried along by the river's flow are deposited at its mouth. The coarser part of the suspended load, mostly sand, is deposited at the river's mouth just as it enters the sea; the finer-grained portion, mostly mud, drifts farther offshore where it slowly settles out of suspension. As shown in Fig. 2-20, the accumulating delta will be laced with long, linear distributary sands that mark successive positions of the river channel on the delta. These sands pass laterally into silts and clays away from the river channel. Levees will also be present, recording overbank flooding. A marsh or swamp will form where the delta builds up to mean sea level.

Over time the river channel migrates back and forth across the top of its delta so that a vertical sequence of deltaic facies builds up: offshore, marine prodelta muds to distributary channel sands to natural levee and back swamp muds (Fig. 2-21). When the river is not actively building one part of its delta (constructional phase), the sea will usually attack and erode it (destructional phase). The waves and currents rework the deltaic sediments, winnowing out the finer-grained materials and leaving behind the coarse sands, creating a *chenier plain.*

Fossils are fairly common in deltaic facies. Shells and burrows of bottom-dwelling invertebrates are found in the marine muds of the delta front and prodelta facies. Vertebrate remains and plant fragments occur in the deltaic swamp facies.

Clastic tidal-flat facies. Coastal waves and currents rework and redistribute clastic sediments along the shore line. In regions where there is a good range in tidal level, about one meter or more, three distinct sedimentary environments form (Fig. 2-22). Above mean high tide is the *salt marsh* with salt-tolerant vegetation, tidal creeks, and occasional brackish ponds. Tidal flats lie between mean high and low water. Marine animals and plants adapted to frequent subaerial exposure flourish here by either burrowing into the tidal-flat sands and muds or by having the ability to withstand desiccation. The lower part of the flat is crossed by gullies eroded by the ebbing tides. The subtidal environment occurs below mean low-tide level and includes not only the organisms and sediments just offshore from the flats, but also those within the gullies and creeks that remain constantly submerged, whatever the tidal fluctuation.

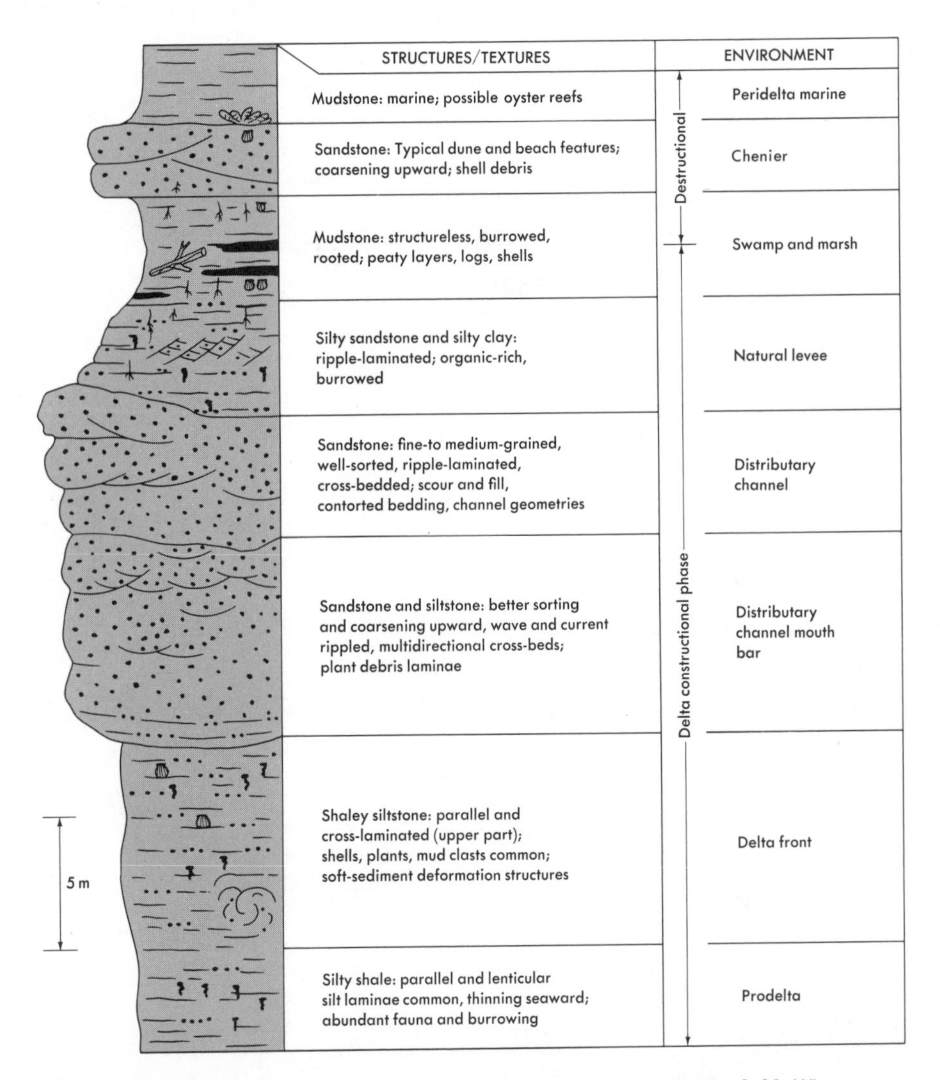

FIG. 2-21 Generalized vertical sequence of deltaic facies. Compare with Fig. 2-20. What causes the systematic changes in average grain size and sedimentary structures going upward in the sequence fron the prodeltaic environment to the chenier sands and bay muds at the top?

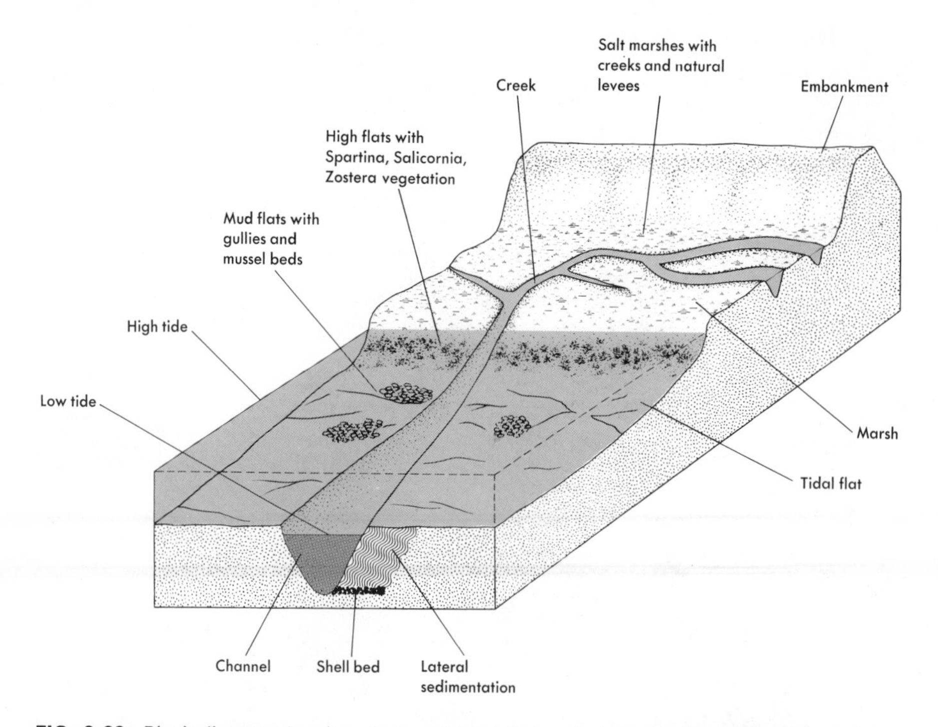

FIG. 2-22 Block diagram showing clastic tidal-flat facies. Periodic flooding and ebbing of the tide moves sediment and water back and forth across the tidal flat. Burrowed muds accumulate in the salt marsh and high part of the tidal flat, while cross-stratified sands are deposited in the lower part of the tidal flat and in the channels and gullies draining the marsh and flat. Migration of the gullies across the flat causes "lateral" sedimentation in the gully; settling out of suspended mud at slack high water results in "vertical" sedimentation on the high flat and salt marsh.

STRUCTURES/TEXTURES		ENVIRONMENT
Mudstone: organic-rich, rooted		Salt marsh
		High water
Mudstone: bioturbated, algal mats, mudcracked	Tidal flat	High mud flat
Sandstone and mudstone: sand decreases upward; burrowing organisms	Tidal flat	Mid flat
Sandstone: herringbone cross-bedding, current and interference ripples	Tidal flat	Lower sand flat
		Low water
Sandstone and mudstone: ripple marks		Subtidal
Conglomerate: mud clasts and shells		Tidal channel lag

1 m

FIG. 2-23 Generalized vertical sequence of clastic tidal-flat sedimentation. Compare with previous figure of tidal-flat environments and account for the changes in fossils, sedimentary structures, and sediment grain size.

Reworked shells and mud clasts accumulate in the tidal channels and gullies; depending on the particular environmental conditions, there may also be beds of attached molluscs (mussels or oysters) living there too. As the channels and gullies migrate across the tidal flat and marsh, these sediments and organisms will move with them. The tidal-flat sediments themselves become finer-grained higher up on the flat, owing to the decreased strength of the flooding and ebbing tidal waters. The lower flat sands display good cross-stratification whose dip reverses itself as the tidal flow changes direction. Higher flat muds show disrupted bedding and stratification owing to the burrowing of tidal-flat organisms. The salt-marsh deposits are burrowed muds with abundant plant remains and root casts. The vertical sequence which we have just described is shown in Fig. 2-23. As with all the facies sequences we have discussed, the tidal-flat sequence will recur repeatedly if the area continues to subside and accumulate sediments, and if the depositional environments remain the same.

Carbonate tidal-flat facies. Along coasts in the warm, low latitudes where little or no clastic sedimentation occurs, there are well-developed carbonate tidal-flat facies. The sediments are calcareous—calcite or aragonite—because they come from the various shelly invertebrates and calcareous algae living there. Evaporation rates tend to be high, so that the sediments are often soon altered and cemented after deposition, owing to solution, precipitation, and replacement of mineral matter from sea water within the sediments.

As shown in Fig. 2-24, carbonate tidal flats, like clastic tidal-flat environments, have a salt marsh or swamp above normal high tide; mud and sand on the flat between high and low tide (but the mud and sand are, as noted, calcareous rather than siliceous); and subtidal, marine sediments below low tide.

Carbonate tidal-flat sediments are characterized by the abundance of algal stromatolites that usually show some evidence of subaerial desiccation, mostly polygonal mud cracks. Desiccation and early cementation of these sediments turn them into rather dense, indurated materials that produce clasts and chips when eroded by strong tides or storm-generated waves. Thus, scour surfaces with limestone-pebble conglomerates are commonly seen in these facies. Winnowed shell beds of marine organisms living in the subtidal environment can be deposited on the high tidal flats by strong tides or storm waves. Figure 2-25 shows the typical sequence of sediments seen in a carbonate tidal-flat environment, starting with the nearshore subtidal deposits and going up through the tidal-flat and supratidal (above normal high tide) sediments.

Reef facies. In the previous section we discussed how reefs can flourish and alter the surrounding patterns of sedimentation. We also indicated how the core of the reef may form a barrier that separates a shallow, quieter water lagoon in the back reef from the deeper, open-sea environment in front of the reef (refer back to Fig. 2-15). We can now ask what sort of vertical sequence will form if the area continues to subside and the reef continues to grow faster than the rate of subsidence—a fairly common geologic circumstance.

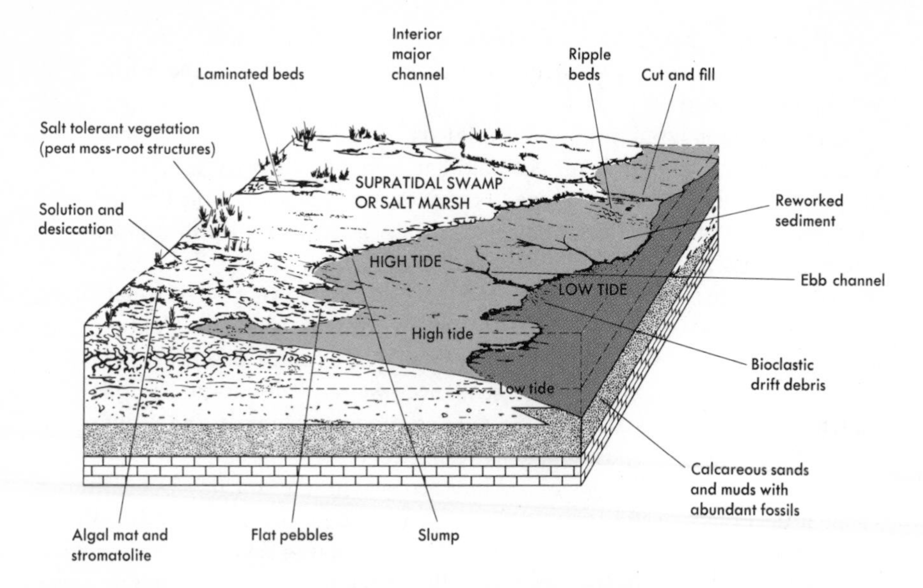

FIG. 2-24 Block diagram of carbonate tidal-flat facies. The horizontal dimension depends on the tidal range and the dip of the depositional surface, varying on the average from tens to thousands of meters. Early cementation, solution/precipitation, and replacement of the calcareous sediments results in their lithification and diagenesis. Subsequent erosion and reworking of the sediment produce clasts that are redeposited, often as limestone-pebble conglomerates along scour surfaces. Algal mats in the tidal flat and supratidal marsh trap and bind sediment, building algal stromatolites whose sizes and shapes depend on strength and frequency of water movement as well as availability of sediment that can be incorporated into the mats.

0.5 m

STRUCTURES/TEXTURES	ENVIRONMENT
Dolomite and evaporites: bedded and nodular; eolianites, stromatolites, mudcracks, and other evidence of desiccation	Supratidal
	Mean high water
Algae: stromatolites polygonally cracked at top and filled with eolian clastics; scour surfaces with limestone-pebble conglomerates	Tidal flat
	Mean low water
Lime sands and muds; abundant shelly fossils; burrows; pellets	Subtidal

FIG. 2-25 Vertical sequence of facies built up in a carbonate tidal-flat environment. Refer to Fig. 2-24 and relate major features of each facies to the individual depositional environments. How does this tidal-flat sequence differ from that formed in a clastic environment as shown in Fig. 2-23?

Using Walther's Law, we can predict that the reef barrier would build out over the fore-reef slope deposits, which, in turn, would migrate toward open ocean. The back-reef lagoon would migrate over the earlier formed reef-barrier facies. Thus, each facies would cover its lateral equivalent in the seaward direction. The vertical sequence that develops is shown in Fig. 2-26.

FIG. 2-26 Vertical sequence of reef facies, going from deep-water, open-sea sediments up through increasingly shallow water sediments into the back-reef lagoon and carbonate tidal-flat facies. Occasional subaerial exposure of the reef core during slightly lowered sea levels results in solution of the reef limestone. Oolite, or rounded grains of calcium carbonate that precipitate from shallow sea water, cap the reef growth. The back-reef lagoon, although also shallow, is a relatively quiet-water environment, owing to its protection from the full force of the ocean waves by the reef barrier.

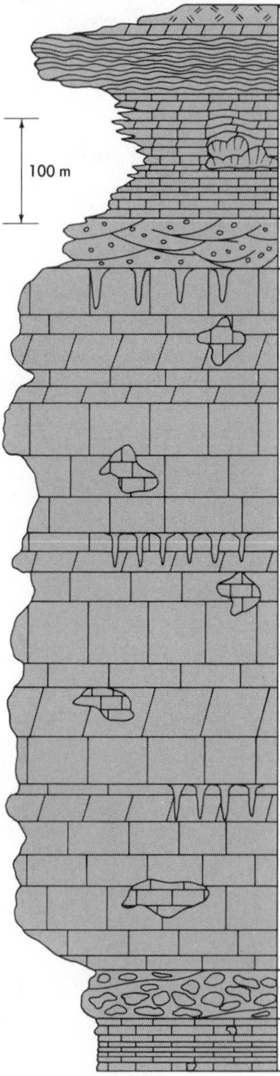

100 m

STRUCTURES/TEXTURES	ENVIRONMENT
Stromatolitic dolomite and evaporites	Tidal flat
Lime sandstone and mudstone: pellets and skeletal debris; dolomitization common; local hexacoral patch reefs	Lagoon with some patch reefs
Oolite: well-sorted and cross-stratified skeletal debris	Inter-reef; back reef
Reef limestone: may be algal, coralline, crinoidal, etc; large primary and solution voids (exaggerated here), which may be filled with later sediment; dolomitization common; subaerial exposure results in solution of reef rock; nonreef building organisms common	Reef
Breccia: angular reef rubble	Forereef slope
Mudstones: dark carbonate, commonly in turbidites; pelagic organisms	Open sea

SUMMARY

Throughout this chapter we have stressed the importance of the local environment in controlling the genesis of a sedimentary rock. Thus, the composition of the source rock and the rate at which it is weathered influence the composition of the sedimentary grains brought to the basin of deposition. The nature of the transporting medium and the conditions existing during and after sediment deposition control the texture and primary structures of the sediments. The postdepositional environment determines the degree and kind of consolidation and lithification of the sediments into rocks. Organisms, too, contribute grains in the form of skeletal debris to the sediments, and the presence and distribution of organisms, in turn, is determined by the local environment. Organisms, moreover, can rework sediments to such an extent that their original textures and structures are significantly altered. Finally, the overall structural and sedimentary framework limits the lateral extent and vertical thickness of a sedimentary unit as well as any facies development within. In short, sediments have complex and changing geologic environments that leave their superposed traces on the resultant sedimentary rock. Careful analysis, however, of the stratigraphic relations, facies, composition, textures, structures, and fossils may reveal the nature and sequence of these individual environments. Walther's Law is especially useful in such analysis because it relates the vertical sequence of rock strata to their horizontal, spatial distribution. Rock sequences, therefore, represent major depositional systems within the stratigraphic record.

three

organisms and environments

In solving ecological problems we are concerned with what animals do . . . as whole, living animals, not as dead animals or a series of parts of animals. We study the circumstances under which they do these things and . . . the limiting factors which prevent them from doing certain other things. (Charles Elton, 1927)

Paleoecologists, like ecologists, wish to understand the conditions of life under which the fossils they study fed, sought shelter, reproduced, and moved about. We must always remember that any particular fossil we are examining—whether an Oliogocene oyster, Triassic amphibian, or Cambrian trilobite—represents a once-living complex biological system, and, as such, was limited in its distribution and abundance by physical, chemical, and biological factors in its environment. Because these environmental factors can only be studied indirectly from the fossils themselves, or from the rocks which enclose them, we need first to learn how modern, living organisms interact with their environment. The knowledge and insight gained from present-day environments and organisms can then be applied to the paleoecology of fossils.

ADAPTIVE RESPONSES OF ORGANISMS

Ordinarily we do not think about the environmental adjustments that animals and plants have made during their evolutionary history. The reason is, of course, obvious. Since human beings are constantly manipulating their environment so that we are able to penetrate and colonize the most hostile surroundings (for example, Antarctica, deserts, and outer space), we have lost an awareness of the strong control that environment usually exerts in limiting the distribution and abundance of animals and plants.

It is a fact that organisms are heavily dependent on the surrounding external world for satisfying their basic needs for living space, food, and mates. And because organisms have varied in their success in adjusting genetically to the local environment, they have had varied success in leaving offspring. Such differential reproductive success results in a gradual change or shift in the genetic background of a species. This change is usually in the direction of increased adaptation to the environment so that with the passage of generations, the species eventually optimizes its total genetic complex with respect to the local environment. As long as the environment remains the same, further changes are selected against and the species remains at its adaptive position. Significant changes in environment, however, will require new adaptations.

There is another, related way in which organisms interact with their environment. The genetic background, or *genotype*, of an individual organism establishes a *range of reaction* of the individual to its environment, and its *particular* expression is influenced by the specific environment in which it finds itself. The resulting, visible product of this interaction between the genotype and the environment is what geneticists term the *phenotype*. Thus, for example, the *potential* size that a given human may attain is controlled by his or her genotype. The *actual* body size that each of us attains is the result of the interaction of our individual genotypes with our individual environments. Differences in the body size of human beings are, therefore, phenotypic expressions of different human genotypes interacting with different environments.

The great diversity that we see in nature is the evolutionary result of many different genetic types invading the wide variety of environments found on our planet. These environments, or *habitats*, are extremely varied and include the prevailing physical and chemical factors, such as temperature, humidity, salinity, content of oxygen and carbon dioxide, and light. Equally important these habitats also include biological factors, such as potential competitors for food and living space, predators and prey, parasites and disease-causing microorganisms, and population density of members of the same species. These habitats may vary from lush, tropical rain forests to hot, arid, treeless plateaus; from cold, dark, organic-rich muds of the deep sea to the warm, well-lighted, agitated waters of a coral reef; from grassy, wide-ranging plains to the dark, moist, and convoluted lining of the mammalian intestine. Within each of these diverse habitats there are a number of *niches* that many kinds of organisms may fill.

Charles Elton, a long-time student of environments and their occupants, has described the habitat as the species' "address," and the niche as the way the species makes its "living" at that address. Continuing the analogy, Elton refers to the typical, rural English village (habitat or address) with its various niches filled by the village vicar, chemist, solicitor, and doctor. Similarly, in the coral-reef habitat there is a multitude of niches ranging from that of the coral polyps sitting in their calcareous colonies and feeding on minute animals suspended in the surrounding sea water to that of burrowing sea cucumbers that eat the bottom sediment for its included organic matter to the predatory barracuda that cruise continuously in search of unwary victims.

The success of an organism in its niche within a particular habitat is measured by its overall adaptation to that niche and the relative number of offspring it leaves as a consequence of that adaptation. The adaptation of an organism expresses itself in myriad ways, including a variety of morphologic, physiologic, behavioral, reproductive, and developmental characteristics and mechanisms necessary to cope with the environment. The pigments in plants, for example, are so colored as to receive particular portions of the available spectrum of sunlight for maximum photosynthetic efficiency. Some flowering plants have evolved a flower structure that, by attracting insects with its nectar, ensures the dusting of an insect with pollen (the male sex cells). When the insect visits another plant of the same species, this pollen is transmitted, thereby fertilizing the second plant. Insect–flower relationships have great reproductive benefits for plants, which are immobile and so cannot seek out a mate. The insects also benefit, of course, by exploiting the food provided by the nectar.

Other types of adaptations are related to the skeletal parts of an organism —wings for flying, fins for swimming, teeth for cutting and chewing food, shell ribbing for burrowing. Consequently, the structure of such parts contains considerable information about the owner's way of life.

Because such skeletal parts are often composed of secreted crystalline materials—bone, shell, or teeth—which are relatively resistant to mechanical disintegration and chemical and bacterial decomposition, they are commonly fossilized. For this reason, then, students of ancient organisms and environments pay particular attention to the structural, or morphological, relations of fossilized skeletal parts in order to make inferences about the habitat and habits of the organism. This emphasis by paleontologists on *adaptive morphology* in fossils is necessary because other adaptive characteristics—whether physiological, behavioral, reproductive, or developmental—are almost always "soft-part" features and hence are rarely preserved in the fossil state. Since protoplasm, tissues, and various organs are completely without preservable hard, crystalline material, they are usually soon disintegrated, decomposed, or eaten after the death of an organism so that burial within sediments and subsequent fossilization is very remote. The loss of information that results from the incomplete preservation of the original living organisms is a principal concern of *taphonomy*, a relatively new subdiscipline of paleontology that studies the processes and events that occur to an organism from the time it dies to the time it becomes a fossil. We will discuss taphonomy in Chapter 4.

It is unfortunately true, therefore, that certain adaptive aspects of ancient organisms will be forever unknown and unknowable. Yet the fossil record is sufficiently adequate to arrive at least at first conclusions regarding the nature of many ancient environments and the roles that various fossil forms played in those environments.

FUNCTIONAL MORPHOLOGY

For an organism to survive and reproduce, it must perform a number of specific functions repeatedly and effectively throughout its life. These functions include, among other things, the searching out of food, shelter, and mates. Food must be gathered and incorporated within the organism where it can be digested and assimilated. This is as true for the elm tree that derives its water and nutrients from the soil through its many thousands of root tips as it is for the sponge that circulates water through its elaborate body-canal system, where millions of small cells filter out the suspended food particles.

Organisms must also establish themselves in an environment where they can survive under the local physical, chemical, and biological conditions. Thus, there must be some mechanism for moving about in the general area and for "settling in" once a favorable site is located. For example, many marine invertebrates, although as adults they may be sedentary, bottom-dwelling forms, have a free-floating period in their early life, or *larval stage*, during which the larvae are dispersed far and wide by the ocean currents. After an interval of time ranging from hours to weeks, the larvae settle to the sea floor and undergo a metamorphosis to a miniature adult. During their free-floating stage, the larvae are usually equipped with some specific morphologic character, such as tufts and bands of cilia, that aid in keeping them suspended in the water and in generating small currents around the mouth for filter-feeding. After metamorphosis, other morphologic structures develop that enable the organism to burrow into the sediment (the foot of some clams), to crawl along the sediment (the various appendages of a lobster), or to secrete a stony, external skeleton that is fixed directly to the sea floor (the calcium carbonate-secreting tissue of certain corals).

Survival of the individual, important as that is, does not necessarily ensure the life of the species. Therefore, organisms must reproduce themselves if the race is to continue. Here, too, one can discover a great variety of morphologic features—as well as physiologic and behavioral mechanisms—that are adapted for the production and fertilization of sex cells, or *gametes*, and, in some instances, for care of the young. The songs, plumage, and nesting habits of birds are all part of an elaborate reproductive complex to attract mates, effect fertilization, and care for the eggs and hatched young.

So we see, then, that organisms have a variety of functions to perform including feeding, locomotion, and reproduction if they and their species are to survive. Now, many of these functions are closely correlated to specific, hard skeletal parts of the organism. If such hard parts are found preserved in the fossil state, then it should be possible to infer the function that such a part played during the life of its owner. This, in turn, should tell us something of the environment where that organism lived.

For example, the skeletal appendages of various vertebrates are specifically adapted—through a long period of natural selection and evolution, of course—

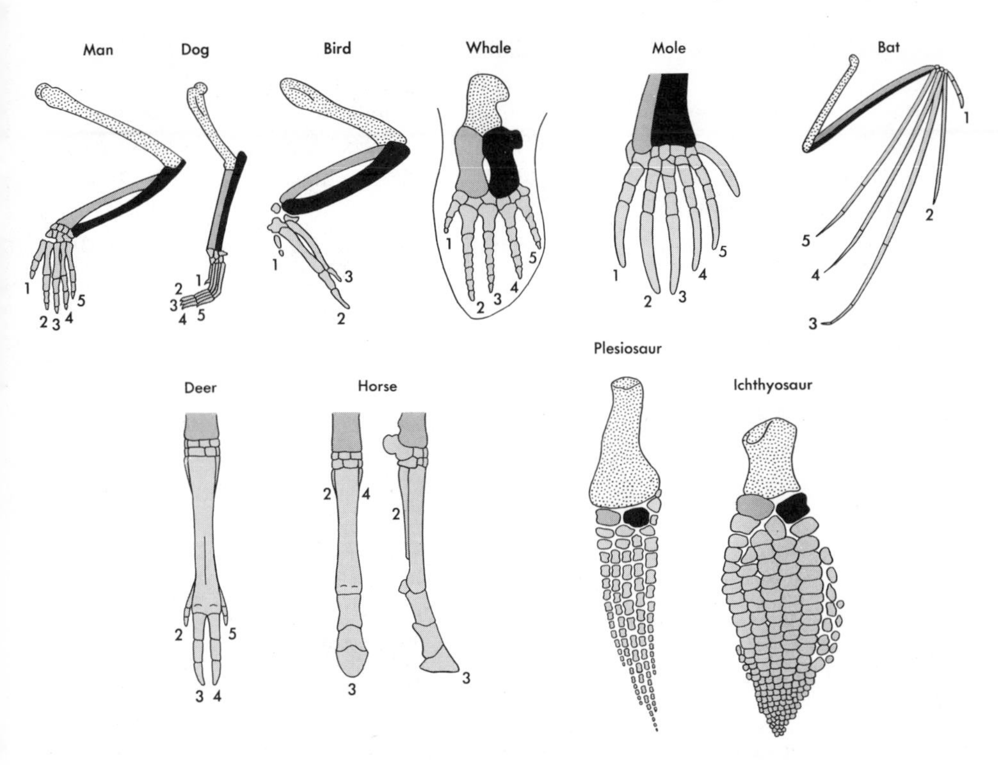

FIG. 3-1 Structure of various vertebrate forelimbs showing the arrangement of homologous bones. Each limb is specifically adapted for a particular way of life, whether burrowing (mole), flying (bat, bird), swimming (whale, ichthyosaur, plesiosaur), running (dog, deer, horse), or manipulating objects and tools (human). Various forelimb bones include: humerus, stippled; radius, dark gray; ulna, black; metacarpals (wrist bones) and phalanges (finger bones), light gray; numbers refer to homologous digits (fingers).

for locomotion in water, on land, or in the air (Fig. 3-1). Fish have short and broad appendages with a relatively large surface area for moving through the dense medium of water. Land-dwelling vertebrates have longer and narrower limbs for supporting the body in walking and running. In the case of those groups that have returned to the sea (ichthyosauran reptiles and cetacean mammals), the limbs have been modified once more and are stubbier and more spatulate than before. Flying vertebrates, the birds and bats, have a limb that, when covered with feathers or a skin membrane, forms an aerodynamically stable structure.

Another example, again from the vertebrates, of the close correlation between the form and function of a morphologic character is provided by mammalian teeth. Every species of mammal, living or extinct, has a unique set of dental characters that are directly related to its diet.

The basic dental formula in mammals consists of 44 teeth that can be subdivided into four sets of 11 homologous teeth: left upper, left lower, right upper, and right lower. In the less-specialized mammals these 11 teeth are, progressing from front to back, three *incisors* for nipping, one *canine* for tearing or stabbing, four *premolars* for cutting, and three *molars* for chewing or grinding. The differences in the shape and size of these teeth reflect a division of masticatory labor among the teeth in biting, cutting, and chewing the food (for three widely varying examples, see Fig. 3-2).

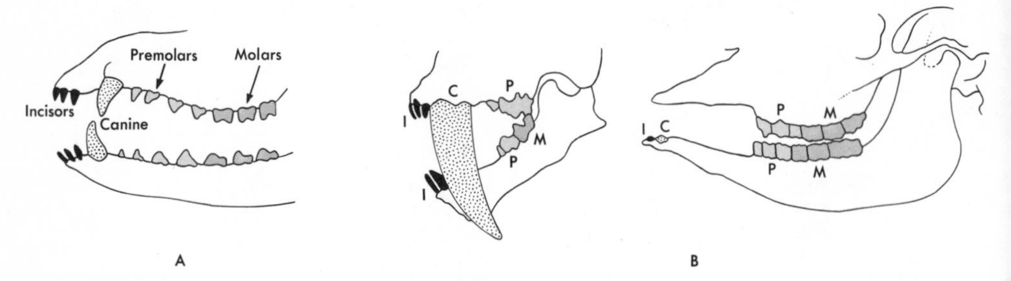

FIG. 3-2 (A) Dentition of a relatively unspecialized mammal showing three incisors (nipping and biting), one canine (stabbing and tearing), four premolars (cutting), and three molars (grinding and chewing) on each half of the upper and lower jaw. (B) Two highly specialized mammals with dentition which is considerably different from that illustrated in (A) : the extinct Pleistocene saber-tooth cat on the left and the modern cow on the right. The saber-tooth cat was a predaceous carnivore equipped with large, stabbing canines and modified upper premolars and lower molars ("carnassial teeth") for cutting up its freshly caught meal. The herbivorous cow has lost its upper incisors and canines; it crops its food of grasses and small plants with the projecting lower incisors that press against a tough, horny pad on the upper jaw. Since a cow's premolars are modified into molar-like teeth, the animal is thus provided with a battery of grinding teeth for thoroughly grinding its food.

Many mammals deviate from this general plan because of specialized diets. The carnivores have particularly well-developed canines for stabbing as well as modified first lower molars and last upper premolars which have become sharp crested shearing blades for cutting—two adaptations well-suited for the eating of freshly killed meat. In contrast, the vegetarian herbivores such as cattle have lost their canines while their premolars have been modified into molar-like teeth suitable for grinding and chewing tough vegetable matter.

Because mammal teeth are composed largely of crystalline calcium phosphate, they are mechanically strong and chemically stable and so are commonly found as fossils in sedimentary rocks. Furthermore, there is a close interdependence between dentition, diet, and way of life. For these two reasons, then, students of mammalian evolution have been able to document and unravel the complex evolutionary history of mammals for the last 100 million years and more, basing much of their interpretation on fossilized teeth.

It is not, however, always easy to demonstrate the functional significance of a particular hard part, principally because many fossil organisms are extinct and lack living, flesh-and-blood representatives that can be observed in the field and dissected in the laboratory. It is often impossible to establish by analogy what function a given hard part served during the life of its owner. Such an example of well-known but poorly understood hard-part morphology is provided by an extinct group of shell-bearing protozoans called *Fusulinids*, which lived some 225 to 325 million years ago during the late Paleozoic Era. During the 100 million years of their existence, the fusulinids underwent considerable evolution, much of which is recorded in various morphologic features of the small, calcareous shell (or test) that they secreted. Some of these changes included overall increase in size, variation in test shape, rotation of the axis of coiling from the shortest diameter to the longest diameter, amount and position of secondary calcium carbonate deposits within the axial portion of the test, and an increase in the fluting of the partitions that internally subdivided the shell. The evolution of these various characters resulted in the rise within this single family of protozoans of some 40 to 50 genera with 1,000 species. And yet, despite the precision with which these evolutionary trends in different morphologic characters can be documented, the functional and adaptive significance of these characters is not at all clear.

In short, therefore, various morphologic aspects of fossil organisms often reveal the habitat and habits of the organism itself as well as contribute information about the ancient environment where that organism lived. But the code relating form to function has not been solved for many, perhaps even the majority, of fossil organisms, particularly those that are extinct and are without a closely related living descendant.

SOME IMPORTANT ENVIRONMENTAL FACTORS: PHYSICAL AND CHEMICAL

Owing to the specific adaptations organisms have made to a given habitat during their evolutionary history, their distribution and abundance in nature, obviously, are determined by the various environmental conditions that make the habitat what it is. These conditions include a complex array of physical, chemical, and biological phenomena that are not only important by themselves but also in their mutual interaction in limiting the distribution and abundance of a specific group of animals or plants. We will review some of the principal ecologic factors that exert an influence on various kinds of organisms. This review, besides being somewhat general and necessarily incomplete, will treat these individual ecologic factors as if they were acting independently; but remember that these factors—as well as others omitted from our discussion—are usually interacting with one another, and that the overall impact of the local environment is greater than the sum of its individual aspects.

Temperature

One of the important factors of any environment is the *temperature* of the surrounding medium, whether water, air or soil. The reason is that the temperature of an organism's surroundings strongly influences its internal temperature in most instances—the most obvious exceptions being birds and mammals, which maintain relatively precise internal temperatures irrespective of the external temperature (within limits, of course).

As the internal temperature of an organism varies—and it varies more or less as the external temperature varies—reaction rates of many physiological processes will also vary because these reactions are mainly chemical in nature and are, therefore, speeded up or slowed down with an increase or decrease in the local external temperature. As a result, rates of metabolism, development, growth, and reproduction will ordinarily fluctuate with temperature changes. This change of reaction rates with temperature change is summarized by *van't Hoff's rule*, which states that for every 10°C of temperature increase, organic reaction rates increase by a factor of 2 (probably more accurately by a factor of 1 to 6). Although this rule cannot be universally applied because of many exceptions, it does suggest the order of magnitude of temperature control on many organisms (Figs. 3-3 and 3-4).

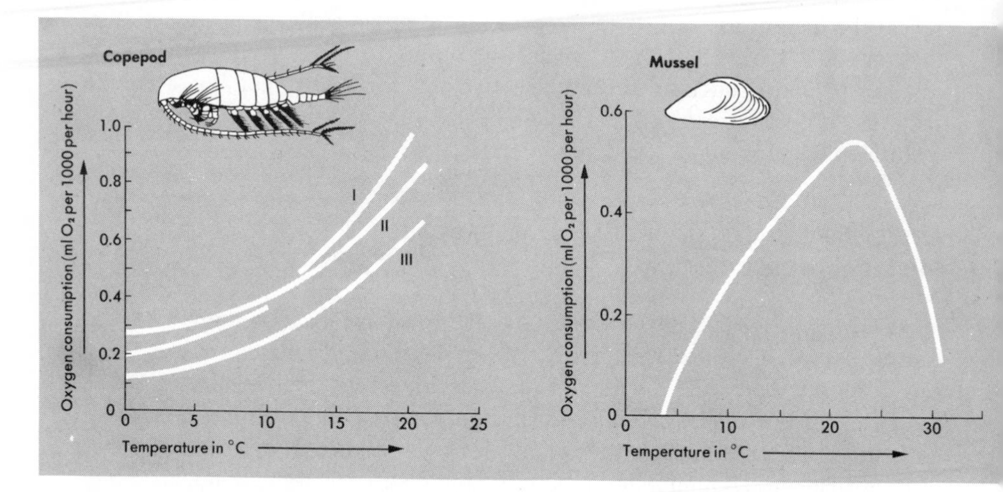

FIG. 3-3 Graph on left shows rapidly increasing respiration rates in male (I), female (II), and juvenile (III) copepods, *Calanus finmarchius*, as temperature of the sea water increases. Graph on right shows similar respiratory increase in the common mussel, *Mytilus edulis*. Notice, however, that above 22°C, respiration rates decline sharply owing to the general debilitating effect of higher temperatures on these organisms. (From H. B. Moore, 1958.)

Besides these direct influences on physiological reaction rates, temperature variations have other, less direct, effects that may be of equal significance. For example, the solubility of various solids and gases that are critical in supporting life, such as nitrates, phosphates, oxygen, and carbon dioxide, vary with temperature: the solids are more soluble, the gases less soluble with an increase in temperature. The availability of these substances to organisms will, therefore, often be related to the surrounding temperature.

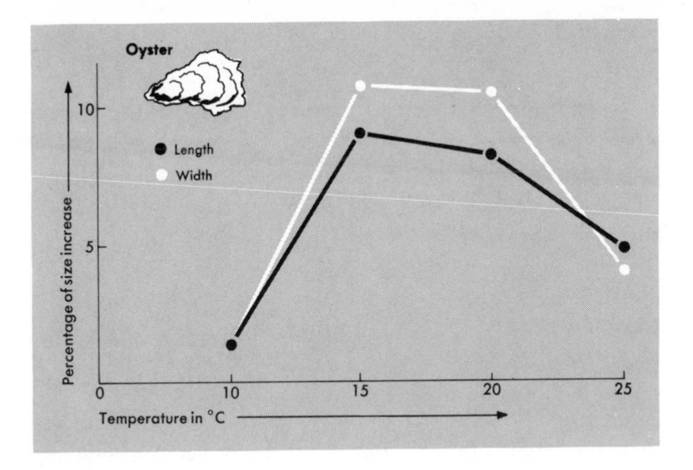

FIG. 3-4 Increasing growth rate (during one month) with temperature increase in the oyster, *Crassostrea virginica*. At 15°C there is an optimum temperature for growth; above this value, growth rates lessen. (After H. B. Moore, 1958.)

Another important way in which temperature influences organisms is through convection of air and water. Periodic temperature variations bring about density differences within air and water masses. Falling temperatures increase the density causing the mass to sink, while rising temperatures will cause it to rise. Such convective movements tend to redistribute and mix the air or water and so increase the circulation of other critical environmental elements contained within the mass, such as nutrients and oxygen. This circulation can, for example, ameliorate an environment that was deteriorating during the stagnant period before the convective circulation.

Oxygen and Carbon Dioxide

Two atmospheric gases, carbon dioxide and oxygen, are especially important to animals and plants. Oxygen is, of course, critical for all organisms because of its role in cellular respiration. Its importance derives from the fact that when various organic molecules such as amino acids, proteins, carbohydrates, and starches, which are either produced by the organism or else obtained from its food, are oxidized, they release their bound-up chemical energy. This energy, which is part of the solar radiant energy originally stored in the organic molecules that are manufactured by plants during photosynthesis, is released in turn in various forms, including those of thermal energy, kinetic energy, and electrical energy, all of which the organism needs to sustain life.

This storage of chemical energy in organic molecules by plants during photosynthesis, and its release by plants and animals during cellular respiration, is summarized by the general equation:

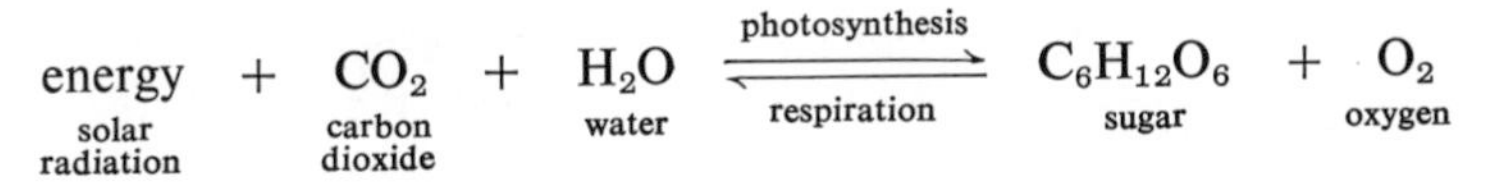

$$\underset{\text{solar radiation}}{\text{energy}} + \underset{\text{carbon dioxide}}{CO_2} + \underset{\text{water}}{H_2O} \underset{\text{respiration}}{\overset{\text{photosynthesis}}{\rightleftharpoons}} \underset{\text{sugar}}{C_6H_{12}O_6} + \underset{\text{oxygen}}{O_2}$$

As this equation states, carbon dioxide is as important to biological processes as oxygen because it is a basic building block for organic matter—either for food or for the formation of new protoplasm.

Although there are some organisms—anaerobic bacteria, in particular—that can liberate energy from organic molecules without oxygen, most organisms need oxygen for cellular respiration. (Some of these, too, can liberate energy from organic molecules without oxygen, but only for limited periods of time because they accumulate an oxygen debt which, like all debts, eventually becomes due.) Oxygen is obtained either from the atmosphere or from solution in water. Oxygen constitutes about 21 percent of the volume of the atmosphere and is normally sufficiently abundant to meet the requirements of all air-breathing organisms. Aquatic organisms, on the other hand, have a smaller and more variable oxygen supply, since the amount of oxygen dissolved in water varies inversely with temperature and salinity of the water. The oxygen content also fluctuates because it is added by plant photosynthesis and subtracted by organic respiration and decomposition (oxidation) of organic matter. As a result, fresh water—lakes, ponds, and streams—and sea water usually contain just several parts per million of dissolved oxygen. Thus, oxygen may be totally depleted in various stagnant aquatic environments (for instance, in depths greater than 200 meters in the Black Sea, or at the bottom of deep, narrow fjords, certain swamps, and deep lakes) which, except for anaerobic bacteria, will be completely devoid of all life.

Carbon dioxide, besides being necessary for photosynthetic organisms, is also ecologically important because it influences the pH, or hydrogen-ion concentration, of natural waters. When atmospheric or oxidative carbon dioxide dissolves in water, it also forms a weak acid, H_2CO_3, which dissociates to form hydrogen ions and bicarbonate ions. The hydrogen-ion concentration, or acidity, of the water, therefore, increases with increasing carbon dioxide in solution. The production of the bicarbonate ion is important in buffering water against sudden changes in acidity or alkalinity.

In the sea, especially, the carbon dioxide content may play a significant role in the solubility of calcium carbonate and, in turn, influence its availability for calcium carbonate-secreting organisms. In the deeper parts of the oceans the waters are relatively cold and under great pressure; they can, therefore, hold more carbon dioxide in solution than the warmer surface waters. At these depths there is solution of the many millions of microscopic, calcareous skeletons secreted by planktonic protozoans in the upper, surface waters. Thus, it happens that in oceanic depths greater than 5,000 meters these skeletons have all virtually dissolved away while falling downward; at shallower depths, however, they become increasingly abundant in the bottom sediments and form, in many places, a fine-grained calcareous sediment. The level at which this dissolution occurs is known as the *calcite* ($CaCO_3$) *compensation depth*, or *CCD*. Interestingly, geochemical studies of sea water indicate that this depth ought to be much shallower than it, in fact, is. Apparently, the thin organic coatings on the planktonic tests initially inhibit dissolution at shallower depths, but are gradually removed by microorganisms as they sink, thereby permitting dissolution at greater depths.

Sunlight

Solar radiation, or light, is another important ecologic factor because it is the basic energy source for *all* organic activity. As indicated earlier, plants convert carbon dioxide and water during photosynthesis into various organic molecules—particularly carbohydrates, as well as fatty acids and proteins—which, in turn, provide a food source not only for the plants themselves but also for all animals, which, unlike plants, cannot carry out this conversion. Therefore, it is through the intermediary process of plant photosynthesis that all life ultimately depends on solar radiant energy.

The amount and duration of solar radiation vary greatly throughout the many geographic areas and environments found on the Earth. Because the Earth's axis of daily rotation is inclined to its axis of yearly revolution about the sun, the solar energy received by various parts of the planet changes, thereby creating seasonal as well as day–night cycles. These seasonal variations in solar radiation are greater the farther one goes from the equator. Whereas plants in the tropics receive strong light for about 12 hours of every day of the year, plants in the arctic regions receive radiation that is much less intense and very unevenly distributed. In fact, in very high latitudes there is continuous daylight in the summer and continuous darkness during the winter.

There is also conspicuous variation in the amount of solar energy received by the oceans and land-locked bodies of water. Besides the latitudinal variations, the depth of light penetration is limited and rarely goes beyond about 200 meters below the water's surface. This zone of maximum light penetration is referred to as the *photic zone* (refer back to Fig. 1-3). Not only does the amount of radiation decrease with water depth, but the wavelength, or color, of the light also changes. The less intense, longer wavelength red end of the visible spectrum—as well as the infrared, or heat, portion—is more readily absorbed by the water than is the more intense, shorter wavelength blue end. It is partly for this reason that there is a general change in the pigmentation of marine algae with water depth. Green algae commonly occur in shallow waters, brown algae in deeper waters, and red algae at the greatest depths. The characteristic depths of these various marine algae reflect adaptations in the development of different colored pigments that are *complementary* to the wavelength (or color) of the light available at these different water depths. There may be other adaptations as well, such as increasing the concentration of photosynthetic pigments to compensate for the decrease of radiation of a particular wavelength. Living phytoplankton and bottom-dwelling plants are limited to this photic zone.

In the *aphotic* regions of the oceans or deep lakes where no light is available, animals must depend for their food supply on the rain of organic matter that comes down from the photic zone. Some animals, of course, prey upon other animals that feed on this drifting organic detritus.

Salinity

In aquatic environments the total amount of dissolved solids, or salinity, is an important environmental factor. Terrestrial waters—streams and rivers, ponds and lakes—differ from oceanic waters in that they are less saline, their salinity is more variable, and they generally contain other proportions of salts. Thus, although the oceans usually contain 3 to 4 percent of salts in solution, river waters vary from less than 0.001 to almost 1 percent (varying, therefore, by a factor of 1,000) and average about 0.01 percent. The bulk (about 99 percent) of the substances in solution in sea water are, in decreasing abundance, chloride, sodium, magnesium, sulfate, calcium, and potassium ions. In river water, on the other hand, the same percentage of dissolved materials is composed of carbonate, calcium, silica, sulfate, chloride, sodium, magnesium, iron and aluminum oxides, and potassium. In sea water the major dissolved constituent is sodium chloride—hence, the "salty" taste of the oceans—and in river water it is calcium carbonate—which forms the scale in tea kettles.

Note the apparent discrepancy here. If the rivers are carrying mostly calcium carbonate to the oceans, why is sodium chloride the major oceanic salt? Chiefly because many marine organisms whose skeletons are made of calcium carbonate are actively extracting large quantities from sea water. Much of this calcium carbonate is buried in marine sediments as fossil debris. If and when these marine sediments are uplifted and eroded, this bound-up calcium carbonate will be carried back to the sea. Sodium chloride, by contrast, is retained in solution, and although it is added to the oceans at a slower rate than calcium carbonate, it has had a greater net accumulation during geologic time.

The single greatest effect of salinity on aquatic organisms is *osmotic pressure*. The cells of organisms are essentially viscous chemical solutions held together by a membrane which, because its function is to keep the cell contents intact, is only partly permeable, allowing just water and certain ions of small diameter to pass back and forth. This movement of water between cells and the exterior is called *osmosis*. When a cell is in a watery medium with a salinity different from that of the cell fluids, the water pressure is directly proportional to the difference in salinity. If the cell's salinity is less than that of the external medium, then water tends to move out from the cell; if the cell salinity is the greater, then water tends to move into the cell. Unless the osmotic pressure and water flow are regulated by the organism, this process of osmosis will result in either the desiccation of the cell through excessive water loss or the flooding of the cell until it eventually bursts.

Most organisms have evolved osmoregulatory mechanisms either for conserving water or for removing excess fluids; such mechanisms allow the organism to tolerate salinity fluctuations in the environment. However, because of the great differences in quantity and composition of the substances dissolved in terrestrial and marine waters, many organisms that originated in the sea have never been able to succeed in colonizing terrestrial waters, for example, echinoderms and cephalopods (Fig. 3-5). Other groups, such as fish, that are able to inhabit both fresh and sea waters, have radically different osmoregulation (Fig. 3-6).

FIG. 3-5 Salinities in the Baltic Sea decrease markedly from the North Sea towards the Gulf of Finland. Paralleling this salinity decrease is a sharp decline in the numbers of species of marine animals. The changes in salinity are due to the freshening of the Baltic Sea by heavy runoff from the surrounding land. (Data from S. G. Segerstrale, 1957.)

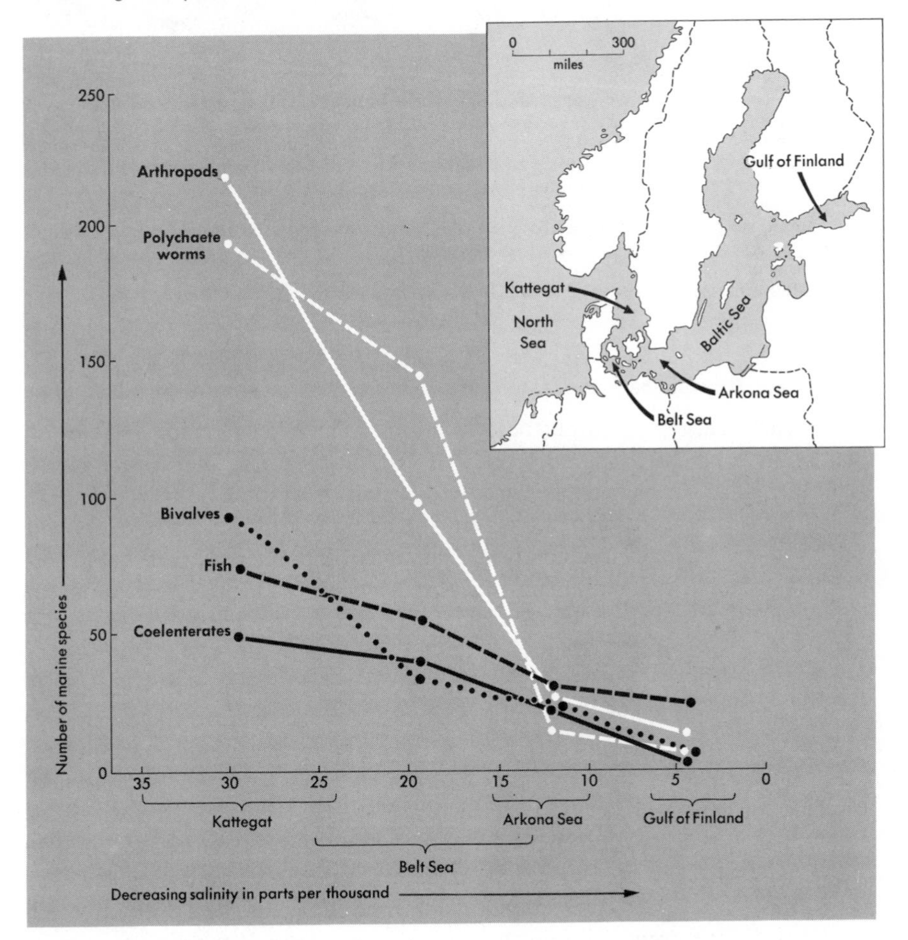

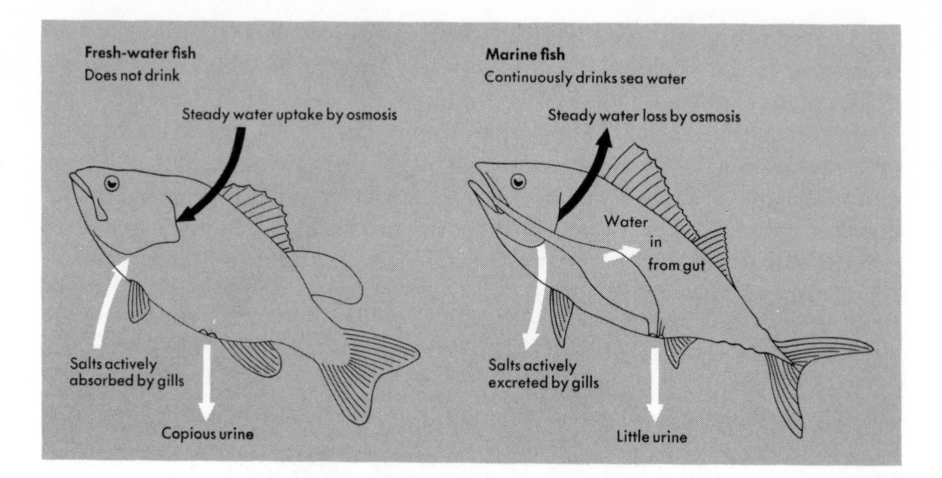

FIG. 3-6 Different mechanisms to maintain proper water balance of the body fluids in freshwater fishes and marine fishes. In freshwater fishes the salt concentration of the cells and tissues is greater than that of the surrounding water. There is a constant absorption therefore, of water by these fish; most of this excess water is removed by a heavy urine flow. To counteract the loss of certain needed salts dissolved in the urine, the kidneys of freshwater fish readsorb a large fraction of these salts from the urine before it is finally excreted. They can also absorb these necessary salts from the water with their gills. Marine fish, on the other hand, have a lower salt concentration in their body fluids relative to the surrounding sea water. To prevent water loss through osmosis they excrete very little urine and drink large quantities of sea water. However, to avoid further upset of their water balance, they secrete through their gills the salts which were dissolved in the sea water that they have drunk. (After Simpson and Beck, 1965.)

Water Turbulence

Water energy, or turbulence, is another important ecologic factor in aquatic environments, especially the shallower parts of the oceans and lakes. Water turbulence helps to ensure an even distribution of food, nutrient elements, oxygen, and carbon dioxide throughout the environment as well as to remove toxic waste products. Consequently, this mixing of the water may prevent a local deterioration of the environment. Water turbulence also disperses the larvae of sedentary invertebrates, thereby helping maintain the areal distribution of these species.

Too turbulent an environment, however, is unfavorable if sedimentary grains are transported along the bottom in such a way that burrowing organisms are uncovered or organisms sitting on or attached to the substrate are buried. If the water energy is too vigorous, bottom dwellers may also be uprooted, dislodged, or overturned (refer back to Fig. 2-8).

Water turbulence determines the size of grains of sediment that accumulate in a given area. The smaller grains settle out in quieter water, whereas the larger grains are carried in suspension or moved along the bottom by traction in more turbulent water. Thus, water energy indirectly influences other characteristics of a sediment related to grain size, such as content of organic matter (an important food source for many organisms), porosity, permeability, and sorting. These characteristics, in turn, control the ease of burrowing into the sediment, water content, and interstitial space for minute sediment-dwellers. (See Figs. 3-8, 3-11, and the discussion in the next section.)

Marine biologists have suggested from time to time that the variations in the shape of certain shallow-water species of snails, sea urchins, and corals are related to differing degrees of wave exposure. For example, the stinging hydrozoan coral, *Millepora*, grows as an encrusting, hummocky colony in very shallow tropical waters. In increasing water depths, where turbulence is reduced, the colonies assume a vertically bladed, labyrinthine shape. Such changes in the gross form of an organism might be useful in interpreting the amount of wave energy or exposure in a fossil environment. Unfortunately, there is not as yet a sufficient body of reliable data relating the gross morphology of an individual organism or colony of organisms to their habitat and, therefore, any such interpretations from the fossil record must be regarded cautiously.

Substrate

Most organisms, even aerial forms, are not suspended in "thin air" but must have a surface, or substrate, upon which they pass their lives feeding, seeking mates, avoiding predators, and resting. In fact, many organisms are substrate-specific, that is, they will search out, and only survive upon, a surface having certain characteristics of background color, food content, firmness, composition, and so on. We know, for example, that many plants will thrive only in soils which have just the right nitrogen content, moisture, and granular texture. Some marine larvae can delay metamorphosis after once settling on the sea floor if the substrate is not suitable; this delay may increase their chances of being dispersed to a more favorable sediment type elsewhere. Certain aquatic species, such as clams and worms, burrow into the substrate for shelter and food.

There are some organisms that prefer a substrate of particular color because they may be virtually invisible there. Some animals, such as the flounder and chameleon, can modify their skin color so that they blend in more harmoniously with their background. The prime advantage of blending in with the substrate is, of course, avoiding or hiding from predators.

An interesting example of the important relation between organism and substrate is provided by a species of moth found in Great Britain. In rural areas today, and throughout the British Isles before the Industrial Revolution, this moth had a "salt and pepper" coloration, that is, dark speckles on a lighter background. Such coloration was obviously adaptive because this particular species of moth spends the daylight hours sitting on light-colored tree trunks and lichen-covered rocks. Against such a background these moths are practically invisible and therefore overlooked by preying birds. With the onset of the Industrial Revolution, however, the trees and rocks in urban areas became very dark, almost black, and many of the light-colored lichens were killed by the fallout of soot, ashes, and other pollutants from factories and mills. Today, except in rural areas where the air is uncontaminated by this industrial fallout, populations of these moths are dominated by a dark variety which is hardly discernible against the soot-covered rocks and trees in areas near the great industrial cities. This dark variety of moth is a mutant strain within the species that has become selectively favored in urban areas of Great Britain since the Industrial Revolution. It is clear, then, that the changing proportions of light and dark varieties of this moth correlate to the color of the background on which they rest during the day (Fig. 3-7).

FIG. 3-7 Light and dark varieties of a British moth showing relative inconspicuousness of each on a sooty tree trunk and on a lichen-covered tree trunk. (Courtesy H. B. D. Kettlewell, Oxford University.)

In the marine realm, the substrate also exerts an influence on organisms. Muddy sediments are usually colonized by burrowing invertebrates that can easily penetrate these soft and watery substrates for shelter and food—either fine-grained organic detritus or bacteria and other microorganisms. Coarser-grained sediments are more difficult to burrow into because the grains are usually well sorted, permitting close packing, and the water content is less. Consequently, such sediments usually have fewer burrowing organisms than do muddier, finer-grained sediments.

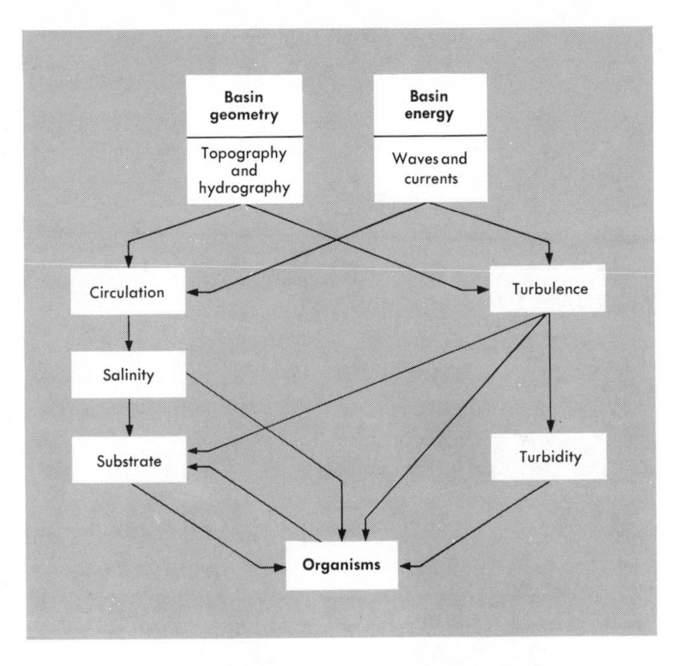

FIG. 3-8 The inferred relationships among the major environmental variables of the Great Bahama Bank in determining the nature of the substrate and the abundance and composition of the bottom-dwelling invertebrates. Note the fundamental dependence of all these ecologic factors upon the geometry and water energy of this shallow, subtropical marine environment. (From N. D. Newell and others, 1959.)

Of all the physical–chemical factors we have discussed, the substrate is especially valuable in understanding *ancient* environments. The reasons for this are as follows:

1. Normally, the substrate of the environment becomes the sedimentary rock matrix with little postdepositional alteration beyond compaction, cementation, and recrystallization (see Chapter 4).
2. Other physical–chemical factors, such as temperature, salinity, and oxygen and nutrient content, are usually *not directly recorded* within the sediments.
3. The substrate is a result of the complex interaction of the topography, hydrography, and current regime of the local sedimentary basin in which the sediments are accumulating and organisms are living.
4. The organisms, particularly the bottom-dwelling forms, are influenced both by the substrate itself and by those other environmental factors associated with a given substrate.

Figure 3-8 illustrates the relationships of the substrate with other ecologic variables as well as the overall control of the sedimentary environment by the topography, hydrography, and water energy of the local basin of sediment accumulation. This diagram is the interpretive result of studies of the dominant ecologic factors responsible for the nature and distribution of sediment types and bottom communities on the Great Bahama Bank. Figure 3-9 shows the

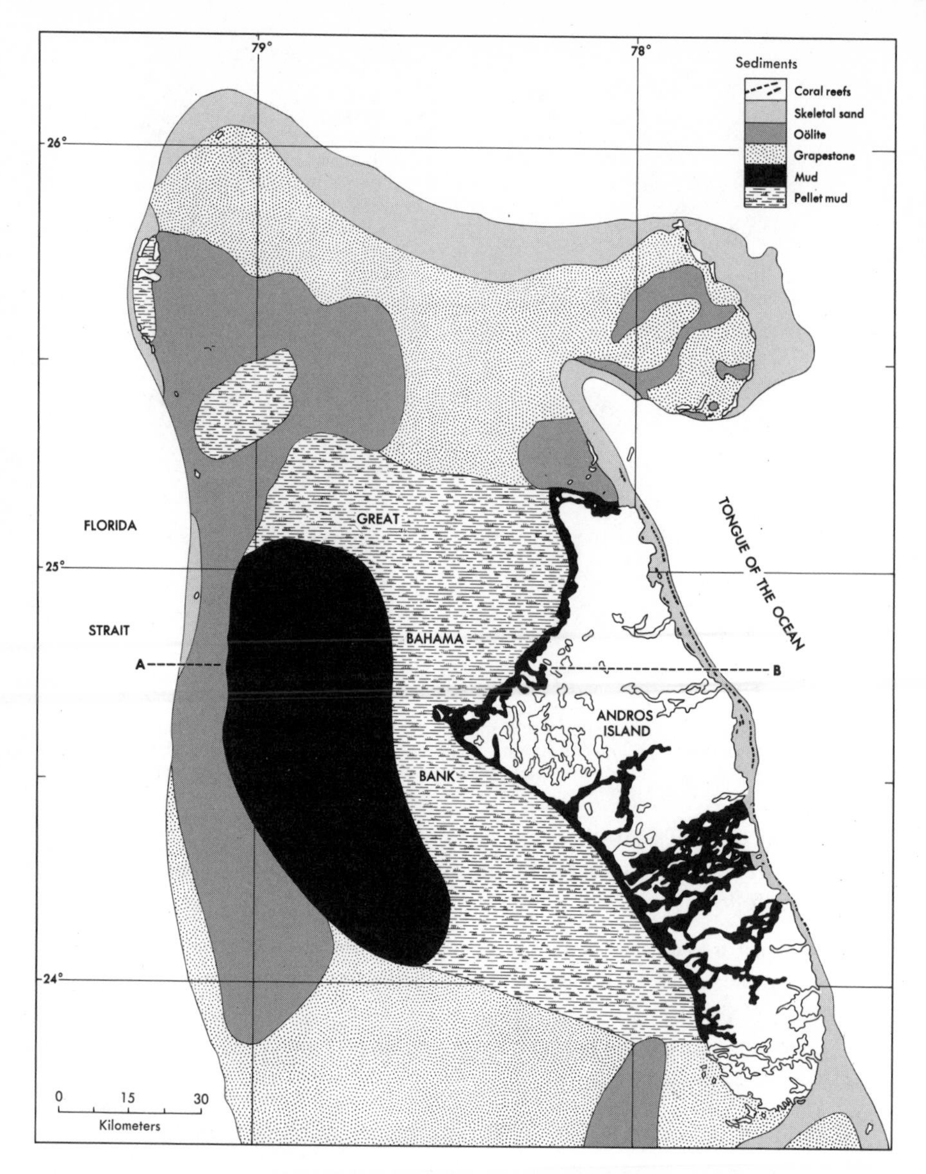

FIG. 3-9 Distribution (above) of major sediment or substrate types and (facing page, top) of bottom-dwelling marine invertebrate communities of the Great Bahama Bank. Note the relatively close coincidence of the distribution of sediment type with the distribution of the bottom communities. At the bottom of the facing page, the cross-section (along line A–B of map above) of the shallowly submerged platform that forms the Bank indicates the importance of salinity, current velocity, and distance from deep, open ocean at the Bank's margin in determining substrate and, by implication, the bottom communities. The *Acropora* community includes corals, calcareous algae, and calcareous protozoans; the *Plexaurid* community includes sea fans, sea whips, snails, scattered corals, sponges, and algae; the *Tivela* community includes a clam, several sea urchins, and marine grasses; the *Strombus* community includes a variety of molluscs and echinoderms; the *Didemnum* community includes a tunicate, green algae, and sponges; the *Cerithidea* community includes a few molluscs, worms, and algae. (After N. D. Newell and others, 1959, and E. G. Purdy, 1964.)

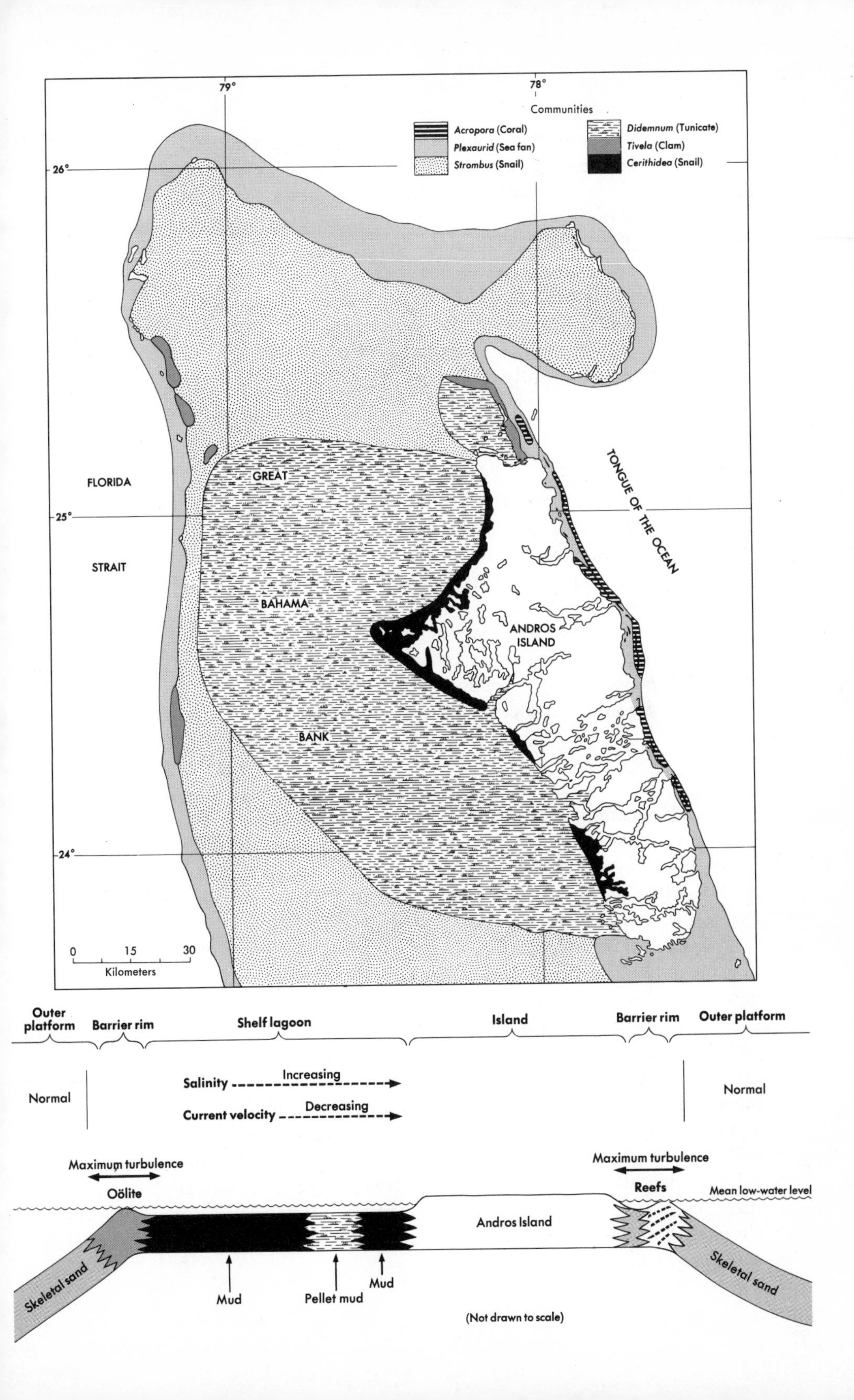

79°
78°
Communities
Acropora (Coral)
Plexaurid (Sea fan)
Strombus (Snail)
Didemnum (Tunicate)
Tivela (Clam)
Cerithidea (Snail)
26°
25°
24°
FLORIDA
STRAIT
GREAT
BAHAMA
BANK
ANDROS
ISLAND
TONGUE OF THE OCEAN
0
15
30
Kilometers
Outer platform
Barrier rim
Shelf lagoon
Island
Barrier rim
Outer platform
Normal
Salinity
Increasing
Current velocity
Decreasing
Normal
Maximum turbulence
Oölite
Maximum turbulence
Reefs
Mean low-water level
Andros Island
Skeletal sand
Mud
Pellet mud
Mud
Skeletal sand
(Not drawn to scale)

close correlation between sediments and bottom communities in this shallow, subtropical, marine environment. The reason for this close correlation is that the composition of the bottom-dwelling invertebrate communities is strongly influenced not only by the substrate itself, but also by the ecologic factors responsible for a particular substrate type. These associated factors are chiefly water turbulence, salinity, and circulation.

This example emphasizes the importance of careful observation and analysis of the rock matrix in which fossils are found—as well as of the fossils themselves—in understanding the major physical–chemical factors operating in an ancient environment.

SOME IMPORTANT ENVIRONMENTAL FACTORS: BIOLOGICAL

Organisms are limited in their abundance and distribution by various biological influences or agents as well as by the physical–chemical factors that we have just considered. These biological factors can be broadly summarized by the feeding relationships found among organisms. These relationships include the food-gathering mechanisms of individual organisms, predator–prey interactions, symbiosis, and food chains and webs. Our discussion here will mainly consider animals since plants make their own food through photosynthesis. The major *feeding types* among animals are:

1. **Herbivores,** which consume vegetal matter. These include such diverse animals as cattle, which graze on grasses; snails that browse on the algal films on rocky intertidal surfaces; various insects that feed on the nectar of many flowering plants; and diverse groups of rodents, which eat grain and nuts.

2. **Carnivores,** which prey on living animals—either herbivores or other carnivores. Besides the obvious carnivorous types such as lions, sharks, and eagles, this category contains such different creatures as starfish, the oyster-drill snail, and the voracious shrew. Also in this group are the carrion feeders, which, like the hyena and vulture, eat the flesh of recently killed animals.

3. **Deposit feeders**, which feed on the organic debris and detritus accumulating on or within the substrate. This group is subdivided into the *selective* deposit feeders, which discriminate between certain kinds of organic matter and avoid as much as possible ingesting the surrounding sediment, and the *nonselective* deposit feeders, which consume the loose substrate for its included organic matter, containing large numbers of unicellular algae, bacteria, and other microorganisms along with organic molecules and tidbits of decomposing animal and plant tissue. Nonselective deposit feeders void as feces the great bulk of this ingested material, digesting only a small fraction of the total volume. Selective deposit feeders include the many scavenging animals, such as certain snails, catfish, and most crabs. Among nonselective deposit feeders are various terrestrial and aquatic annelid worms, sea-cucumbers (holothurian echinoderms), and some clams.

4. **Suspension feeders,** which, by various processes, strain or filter out of water the suspended organic mater or microscopic organisms. Suspension feeders are as diverse as the baleen whale, which sieves out through the long, fibrous filaments growing from its upper jaw the many thousands of small swimming arthropods living in the sea; many species of clams that bathe their gills with a flow of water from which suspended organic matter is gathered up and brought to the mouth by strands of mucus; and the sponges, which pump water through their elaborate canal system lined with special cells for capturing and ingesting fine-grained suspended food particles.

5. **Omnivores,** which are varied feeders employing two or more of the above feeding types, depending on the availability of a particular food. Humans as well as most other primates, are included within this category.

Obviously, the distribution of these various feeding types will depend on the distribution of the requisite food source. Thus, sedentary suspension feeders prefer water that is sufficiently agitated to ensure that organic matter remains in suspension and is continually transported within their feeding range (Fig. 3-10). These animals will be mostly *epifaunal,* that is, they will live on the substrate.

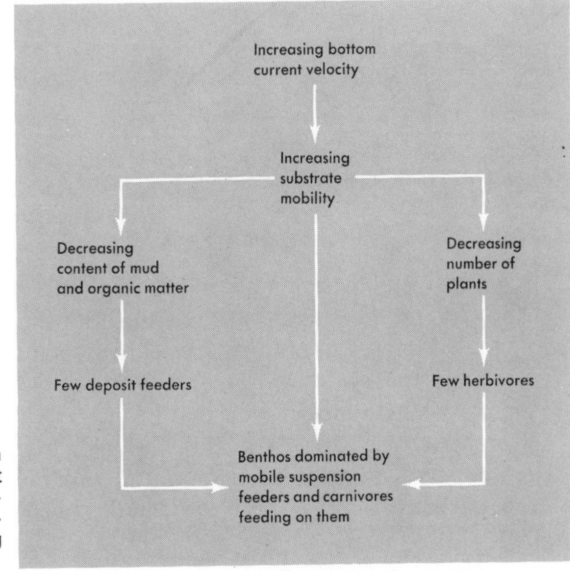

FIG. 3-10 Simplified diagram showing how increasing current velocity along the substrate influences the kind of bottom-dwelling animals, or benthos, living there. (From Purdy, 1964.)

In quiet water, suspended organic matter settles on the substrate. Here, there will be many *infaunal* animals that are deposit feeders burrowing through the sediments for their included food. Figure 3-11 shows the changing proportions of epifaunal suspension feeders and infaunal deposit feeders in Long Island Sound with increases in water depth that control water energy, sediment size, and organic and water content of the sediments.

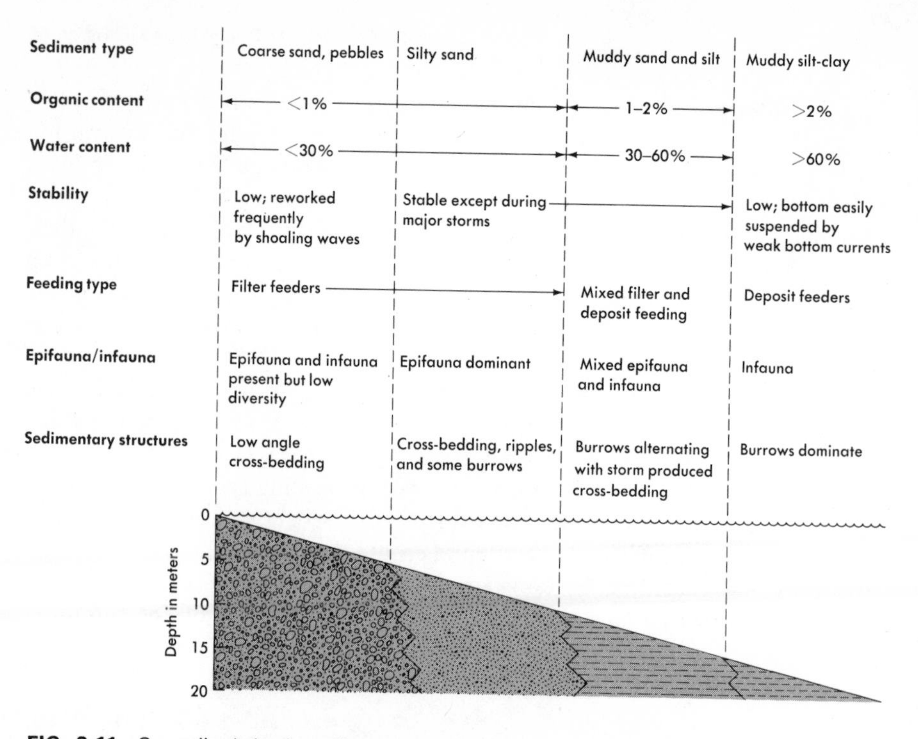

FIG. 3-11 Generalized depth profile on Long Island Sound showing the changes in sediment grain size, organic content, and water content and the resultant control on the bottom-dwelling animals, in terms of how they feed and their relation to the substrate. Note that in shallower water the rate of reworking by organisms is less than that of waves and currents so that the sediments are cross-stratified and rippled. Burrowing in deeper, quieter water disrupts these inorganic structures. (From Rhoads and Berner, 1968.)

Carnivore distribution parallels the distribution of their prey. Of course, carnivores may be rather far-ranging, moving into different, specific areas to find their prey. For example, a mountain lion will descend into several different valleys to seize a variety of herbivores. Still, the overall distribution of mountain lions coincides with that of their prey. Moreover, the abundance of the prey may also control the relative abundance of the predator. Declines in the prey population, either because of too much predation or from some other extrinsic factor, may cause a corresponding decline in the numbers of predators (Fig. 3-12).

Besides predator–prey relationships, there are other close associations between different species. Some of these associations may be so close and constant that they are treated as a separate biological phenomenon—*symbiosis,* which literally means "life together." Symbiotic relationships can be divided into three types: those that provide mutual benefit to the participating species (*mutualism*); those that benefit just one participant with little or no advantage gained by the other species (*commensalism*); and those that benefit one species to the detriment of the other (*parasitism*).

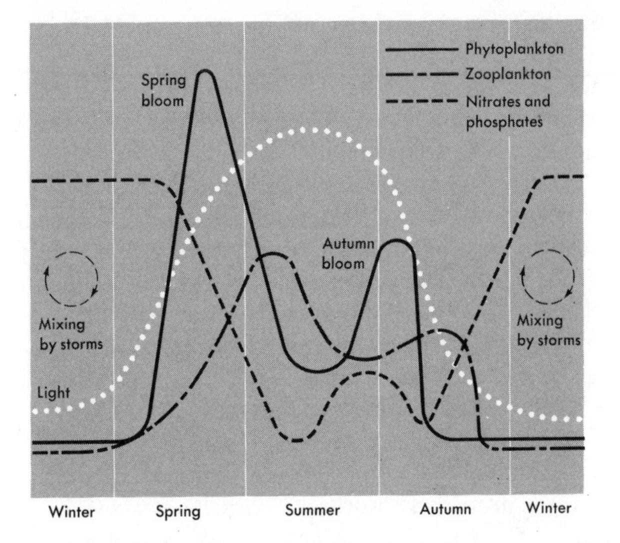

FIG. 3-12 Seasonal fluctuations in light and nutrients (nitrates and phosphates) in shallow waters of the Western Atlantic Ocean control the abundance of microscopic floating plants, or *phytoplankton,* which bloom in the spring and in the fall. The small swimming and floating animals, or *zooplankton,* that graze on the phytoplankton rise and fall with them. This graph shows not only the interconnections between consumer (zooplankton) and producer (phytoplankton), but also the interconnections with other factors in the environment, like light and nutrients, as well as seasonal storms and temperature variations that circulate the nitrates and phosphates into shallow waters. (After Simpson and Beck, 1965.)

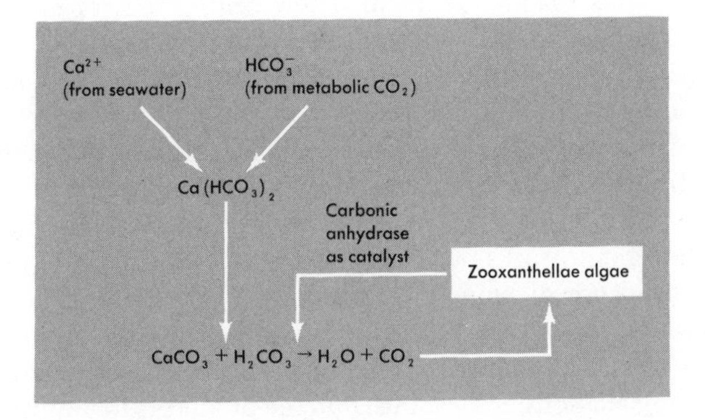

FIG. 3-13 Diagram showing how the production of carbonic anhydrase by zooxanthellae algae facilitates the secretion of calcium carbonate by stony corals. The corals initially secrete calcium bicarbonate, using calcium from sea water and bicarbonate from metabolic carbon dioxide. The calcium bicarbonate, which is precipitated on an organic matrix, subsequently forms calcium carbonate—the mineral aragonite—and carbonic acid (H_2CO_3). Because shallow marine waters are supersaturated with respect to $CaCO_3$, small decreases in H_2CO_3 and CO_2 will greatly increase the formation of skeletal aragonite. Algal production of carbonic anhydrase catalyzes the dissociation of carbonic acid to water and carbon dioxide; algal photosynthesis removes the carbon dioxide. Thus, when reef corals are exposed to daylight, the algae and carbonic anhydrase exert their effects synergistically, and the reef corals can secrete great amounts of skeletal calcium carbonate. (After T. Goreau, 1961.)

An interesting example of mutualism is provided by the stony, reef-building corals and zooxanthellae algae, a diverse group of unicellular plants that live within the coralline polyps. The polyps provide the algae with shelter, metabolic carbon dioxide, and wastes rich in nitrogen. The algae probably give the polyps photosynthetic oxygen and some nutrients as well as remove metabolic wastes. But the most important contribution of the algae—and one that has been unambiguously demonstrated—is an enzyme, carbonic anhydrase, which facilitates the secretion of calcium carbonate skeletons by the polyps, as shown in Fig. 3-13. It appears reasonable that such a relationship would be mutually beneficial for corals and algae. The algae help the corals secrete the great masses of calcium carbonate necessary for reef growth as well as provide food and oxygen, not to mention remove noxious metabolites. The corals, in turn, provide a safe, well-lit environment rich in carbon dioxide and nutrients for the algae.

Commensalism, which etymologically means "feeding together," is illustrated by the marine sand worm, *Clymenella torquata*, commonly found along the eastern shores of North America. This nonselective deposit-feeding polychaete worm builds a tube of mucus-cemented sand grains that projects from the sediment surface downward for several centimeters into the sediment. Living head down in this tube, the worm feeds on the sediment at the lower end of the tube and it discharges feces from the upper end. Commonly, a small, suspension-feeding clam, *Aligena elevata*, attaches itself to the lower opening of the tube. The clam is taking advantage of the improved water circulation generated by the feeding activities of the worm. The water contained in the sediments, even at this shallow depth of a few centimeters, is not usually well circulated and consequently the content of oxygen and nutrients is diminished. By attaching itself to the worm tube, the clam has more oxygen and food available to it than it might otherwise have.

Parasitism is seen in the adult lamprey, which fixes itself directly to various marine and freshwater fish by means of a sucker disk around the mouth. The lamprey is attached to the host continuously, feeding on its soft tissue until the host dies. The lamprey seeks out a new victim and the process is repeated.

The dependence of animals ("consumers") on plants ("producers") is shown very simply in the case of herbivores such as cattle which feed directly on grass. But since not all animals are herbivores, there is often a sequence, or chain, of feeding relationships leading from the producers to ultimate consumer. Each link in this *food chain* supports the next, starting with the photosynthesizing plants, passing through a number of intermediary animals, and finally ending with a particular animal (Fig. 3-14). In any given environment there may be many such food chains, each of which may include some links in the others. The result is a complex interaction—*food web*—among the animals and plants living in this environment (Fig. 3-15).

It should be clear, therefore, that in virtually all natural environments the abundance and distribution of any particular species of organisms are controlled

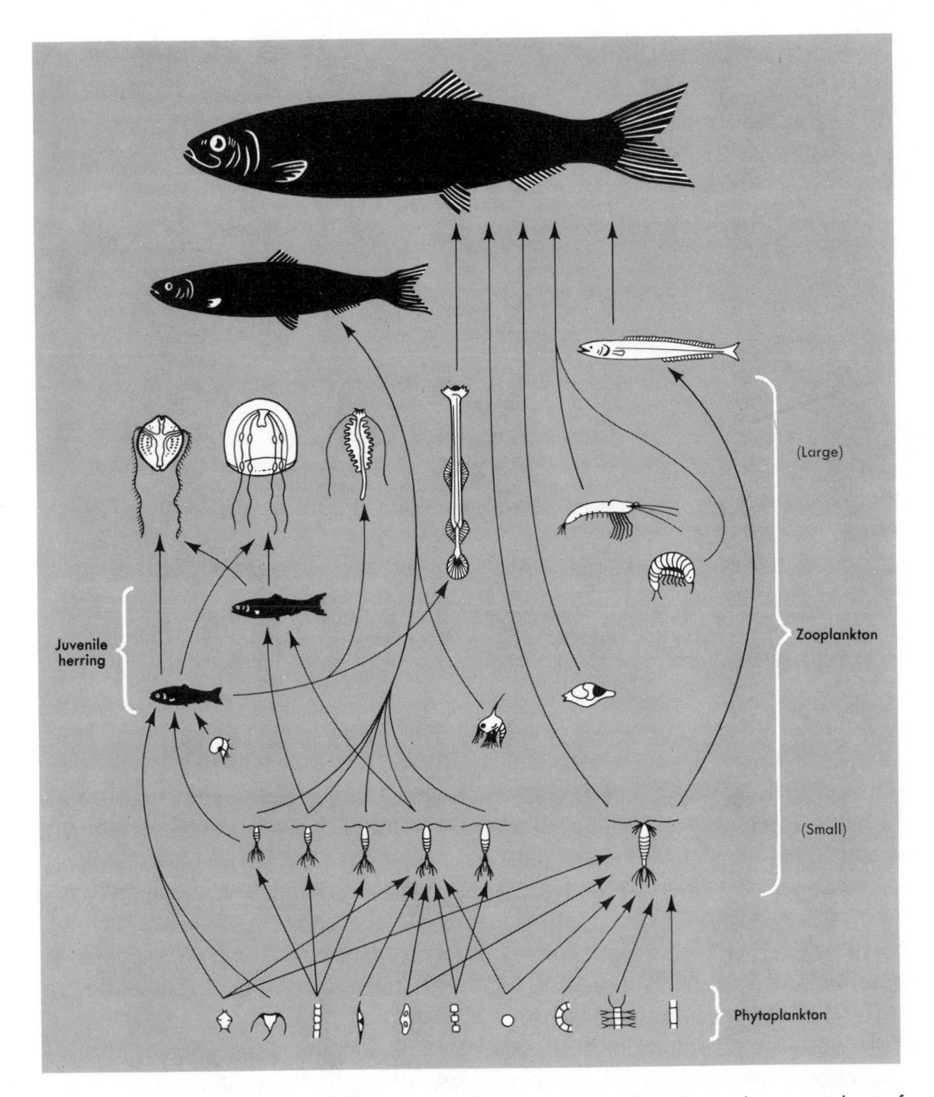

FIG. 3-14 Food chain of the herring, showing the various links from the producer organisms of several different kinds of phytoplankton (small floating plants) through a link of zooplankton (small floating animals) composed of different marine invertebrates, to the final link of the adult herring. Notice that the juvenile and smaller herring feed on the smaller zooplankton. (From Hedgpeth, 1957.)

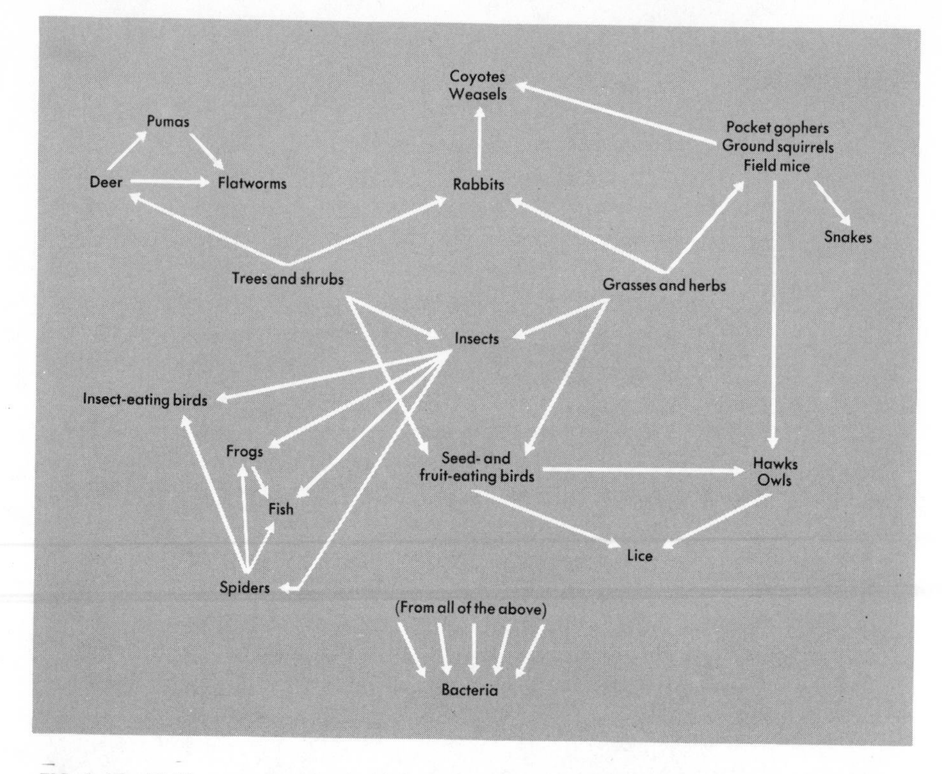

FIG. 3-15 Highly generalized food web in an American woodland community, showing the flow of energy through various feeding relationships. The actual interactions among these organisms and others not included in this simplified diagram are far more complex. This complexity is further compounded when we consider the various physical and chemical factors that influence the abundance and distribution of these different species. (From Simpson and Beck, 1965.)

or influenced by a variety of physical, chemical, and biological factors. As the number of species increases these various interactions among organisms become compounded. Because of these complex interrelationships among organisms with one another and with their inorganic surroundings, it is extremely difficult to isolate any single factor as the most important one. Yet, despite this complexity, it is sometimes possible to define some one ecologic factor as being particularly critical, or limiting, so that most of the others either hinge upon it or are by comparison somewhat less important. This concept of the *limiting factor* is often referred to as "Liebig's Rule," after the nineteenth-century biologist Justus Liebig who first formulated it.

An example of a limiting factor is the amount of available nitrates and phosphates in an aquatic environment, where the total abundance of the plants, and by implication the animals, depends quite closely on these nutrient substances (refer back to Fig. 3-12). Should the availability of these nutrients fall below some minimum level, the plants and animals would perhaps be severely affected. It should be emphasized, however, that recognition of a limiting factor in an environment does not imply that other ecologic factors can be ignored altogether.

ASSEMBLAGES AND COMMUNITIES

So far we have discussed how *individual* organisms are limited in distribution and abundance by physical, chemical, and biologic factors in the environment. If these relationships hold true for single organisms, we can logically conclude that they also hold true for aggregations or assemblages of organisms that live together in a particular place. Consequently, the composition of such assemblages in terms of their component species of plants and animals must reflect the ecologic conditions prevailing where they are found. In fact, this becomes obvious once we consider the different assemblages of organisms that occur across the face of the ecologically diverse Earth.

In this section we want to explore some of the specific ways in which assemblages of organisms are related to the environments in which they occur. We will also be addressing ourselves to the question of why major habitats tend to have characteristic organisms recurring there. For example, coral-reef habitats around the world have rather similar assemblages of organisms, as do tropical deserts or sandy beaches. Moreover, we see recurring associations of fossil animals and plants in the stratigraphic record. To what extent are these merely chance recurrences, and to what extent are they due to ecological interdependance? And if ecologically controlled, can we interpret the ancient environment?

In some instances, the aggregations of different species in nature may be the simple result of overlapping ecologic tolerances of the individual component species with little or no interaction of the species with each other (Fig. 3-16).

FIG. 3-16 Schematic diagram illustrating how two different assemblages of organisms may reflect different overlapping ranges of ecologic tolerances of component species. In both cases the association of the individual species is the result of mutual ecologies with a minimum of species interaction.

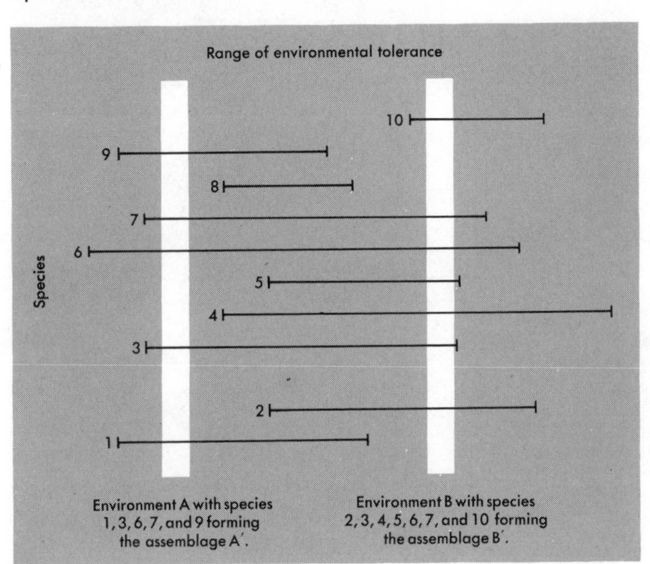

A rough analogy might be the composition of a crowd in the Rose Bowl or at a rock concert, which reflects mutual interest in football or popular music rather than any significant interpersonal dependencies among the spectators.

In direct contrast to such a *statistical association* of species is the concept of an *integrated community* (refer back to Fig. 3-15) in which there is a very high degree of interaction among the component species, especially in feeding and use of space. Such an integrated community can be viewed as a more or less closed ecological system, or *ecosystem*, in which energy and matter are continuously circulated (Fig. 3-17). When examining an ecosystem, we are as interested in identifying the ecologic role of the constituent species as we are in saying what kind of plant or animal it is. Thus, when considering a modern or Silurian coral reef ecosystem we want to know which animals were herbivores and which carnivores, and which plants were the primary producers. (For example, compare Figs. 1-1, 3-15, and 3-17: Can you specify the ecologic roles of the various organisms?)

When studying a particular ecosystem—whether modern or ancient—we want to identify, if possible, the ecologic role of each species in the system. We try to define this role in as much detail as we can. Thus, besides identifying primary producers, herbivores, and carnivores, we also want to know which herbivores ate which plants, predator–prey relations, and any symbiotic asso-

FIG. 3-17 Concept of the ecosystem in which energy and matter are continually cycled through various parts of the system, including primary producers, consumers, scavengers, and decomposers. Notice that different parts of the system can be subdivided further (different kinds of herbivores, for example); note, too, that only the ecologic functions of the system are identified, not the particular species performing that function. Which species does what, of course, depends on the specific ecosystem being considered. (Compare with Fig. 3-15.)

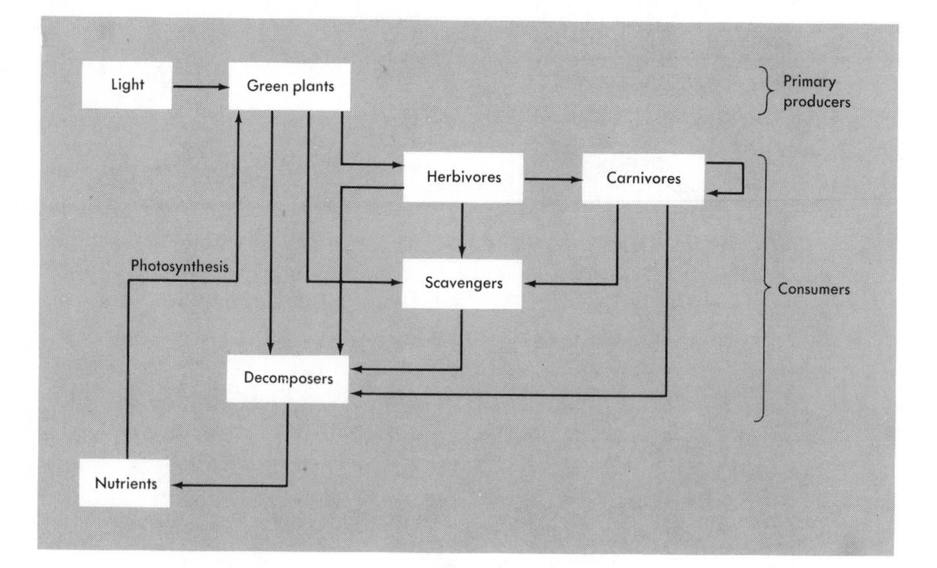

ciations. Naturally, this is more easily done for present-day ecosystems, where we can observe directly the interactions among the species, than it is for fossil ecosystems, where we not only cannot observe directly what's going on, but worse, many of the original species are not even present owing to lack of preservable hard parts.

In paleoecological analysis of fossil marine communities, it is useful to define which species are planktonic (floaters), nektonic (swimmers), and benthonic (bottom-dwelling). This last category can be further subdivided into forms that live on or in the substrate—epifaunal or infaunal—and whether they are mobile or sedentary. Once this is done, we then try to determine how each species fed: carnivore, herbivore, scavengers, suspension feeders, and selective or nonselective deposit feeders. By making this kind of analysis, we have a much firmer idea of how the resources in the environment—food and living space, especially—were utilized. In Chapter 6 we will discuss some specific examples of this type of paleoecological analysis.

Besides local differences in the composition of organic assemblages owing to local environmental influences, there are also some interesting regional or geographic variations in such assemblages. For example, the *diversity* of animal and plant species increases away from the Earth's polar regions toward the equator. Thus, in the boreal climates, forests are usually composed of only a few species of trees such as pines and birches. In contrast, tropical forests may have many hundreds of different plant species. Similar latitudinal diversity gradients have been observed for many other organisms, including birds, snakes, various marine invertebrates, and even for some fossil assemblages (Fig. 3-18). The *numbers of individuals* of a particular species, however, increases with decreases in diversity of species toward the poles. This gradient may merely reflect an increase in food resources and living space with the decline in number of species.

Interpreting diversity gradients is not as easy as observing them. One suggested explanation is that in environments where physical and chemical conditions fluctuate widely, or where the conditions are highly unfavorable regardless of fluctuations, the physiological stresses on the organisms are high, and only a few species have been able to adapt to them (Fig. 3-19). Such communities are referred to as *physically controlled* ones that concentrate their adaptations to either a broad range of environmental fluctuations or very harsh physical conditions. Examples of these communities are found in hypersaline bays, arctic tundra, and tropical deserts. In contrast to physically controlled communities are *biologically accommodated* communities where physical conditions are fairly constant and uniform for long periods of time, and consequently the organisms evolve adaptations geared toward solving biological interactions such as competition for food, predator–prey relationships, and symbiotic associations. Examples of such biologically accommodated communities are coral reefs, tropical rain forests, and deep-sea abyssal plains.

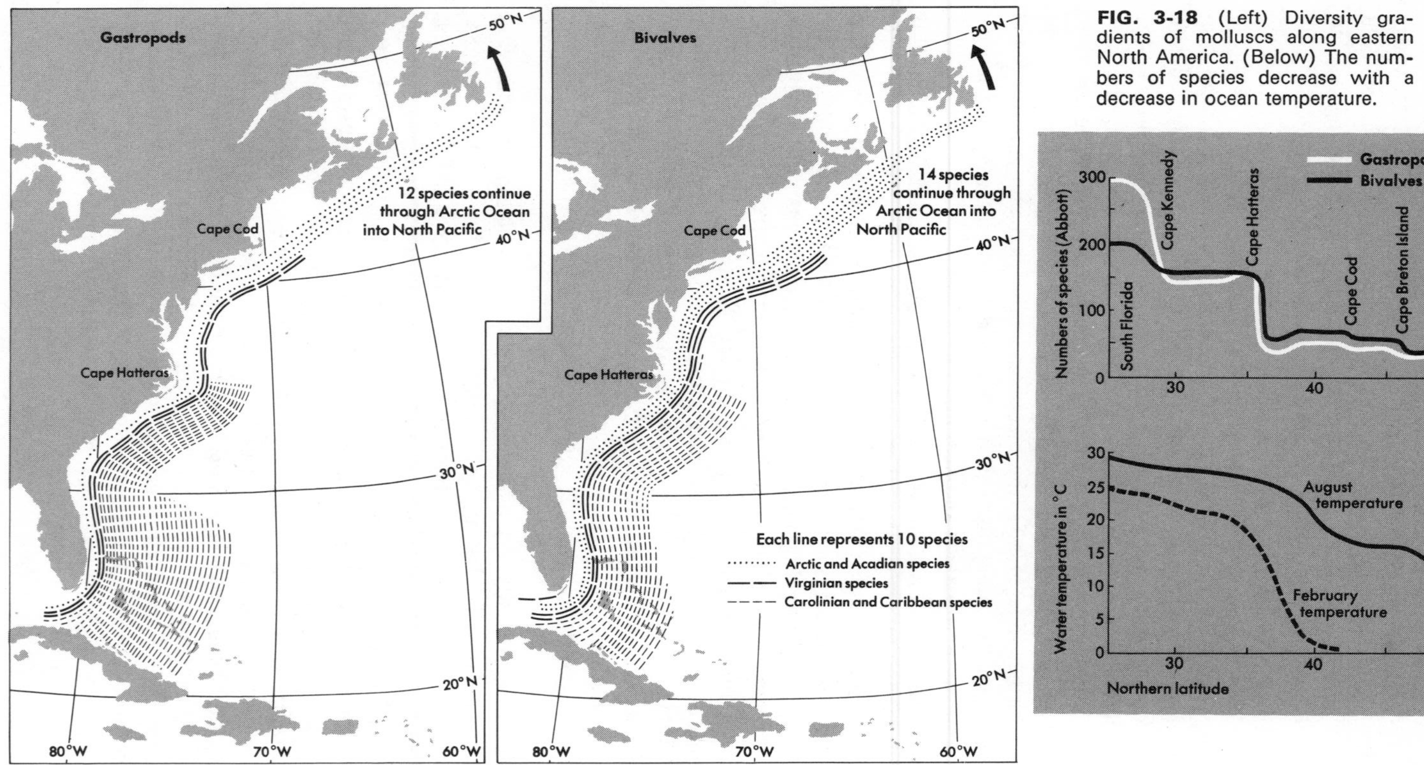

FIG. 3-18 (Left) Diversity gradients of molluscs along eastern North America. (Below) The numbers of species decrease with a decrease in ocean temperature.

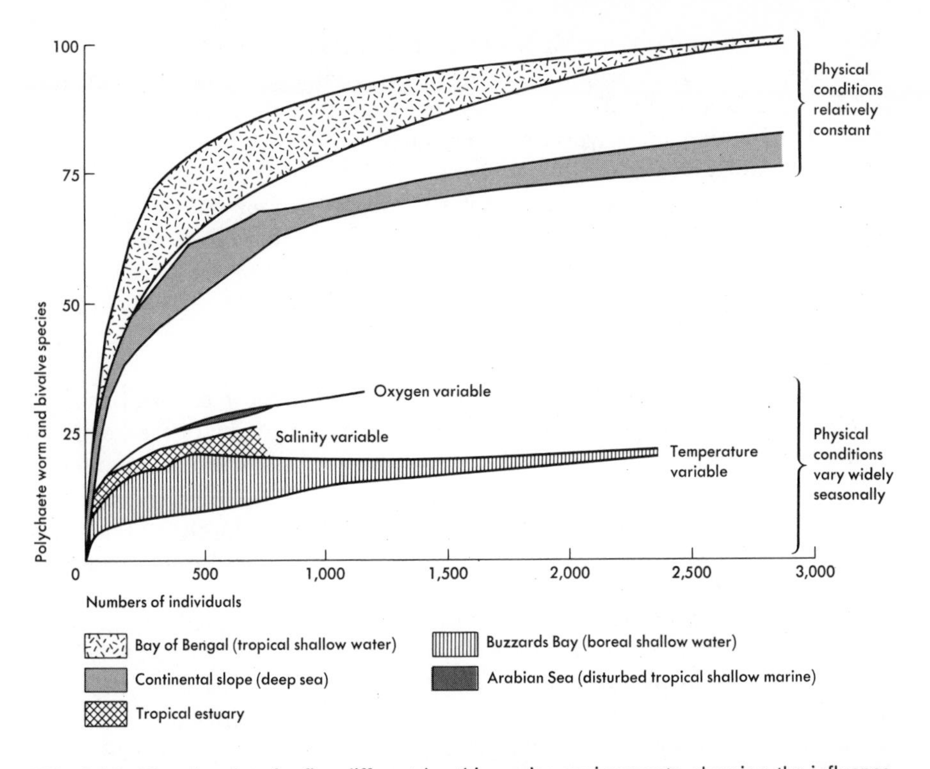

FIG. 3-19 Diversity plots for five different benthic marine environments, showing the influence of varying physical conditions on the numbers of species of marine worms and clams. The shape of each of the curves indicates that as more and more individual animals are observed in an environment, fewer and fewer new species are encountered. In the Bay of Bengal and along the deep sea continental slope, physical conditions are relatively constant; here the benthic communities are biologically accommodated. The other three environments experience wide seasonal fluctuations in either oxygen, salinity, or temperature; here the communities are physically controlled and have much lower diversity. (After Sanders, 1968.)

According to this hypothesis, therefore, global latitudinal or onshore–offshore gradients in diversity and abundance of individuals reflect a gradient from biologically accommodated communities in low latitudes or nearshore to physically accommodated communities in high latitudes or offshore. As Fig. 3-19 shows, however, there can be local physically controlled communities in low latitudes where temperature, salinity, and oxygen fluctuate widely.

In addition to all the various ecologic conditions controlling the character of the Earth's animal and plant assemblages (thereby defining the planet's *biogeography*), there are two other important factors: barriers to dispersal of organisms and history. For example, some organisms may be prevented from occupying a suitable environment because they cannot breach barriers to their dispersal. Such barriers for land dwellers include mountain ranges, deserts, and large bodies of water. Marine organisms, on the other hand, may be prevented from invading a suitable environment because of intervening land masses or widespread oceanic deeps. Thus, the unique character of the flora and fauna of Australia is a result of its being isolated from the rest of the continents for millions of years. Similarly, the shallow-water marine invertebrates in the Indo-Pacific region differ considerably from those on the western shores of the Americas because of the great expanse of deep water of the eastern Pacific Ocean separating these two areas. Few shallow-water, bottom-dwelling marine invertebrates have sufficiently long-lived floating larval stages to survive the trip necessary to go from the western to the eastern Pacific. The lack of abundant islands in the eastern Pacific also inhibits island-hopping across the several thousands of kilometers of water.

Biogeography has another important element, besides ecologic and dispersal factors—the historical element. Because the geography of the Earth has changed through time, the distribution of organisms has correspondingly changed. Consequently, the flora and fauna of any one particular place at a particular time are a result of the existence of earlier organisms in that place (the historical factor) as well as the environmental conditions found there (the ecologic factor).

A revealing example of historical factors determining biogeographic patterns is offered by the South American mammals. In the early Tertiary Period, South America was connected to North America by the Isthmus of Panama and was populated by a variety of primitive mammals. Shortly thereafter, the Isthmus of Panama was submerged under the sea, and South America became isolated from the rest of the world. During this period of isolation the ancestral mammals of early Tertiary time evolved into a number of different groups, including various types of terrestrial herbivores and carnivores. The isolation of South America was not complete, however, because a few animals were introduced around the middle of the Tertiary Period. These forms included certain species of New World monkeys and rodents which, apparently, island-hopped from North and Central America to the South American mainland.

Toward the close of the Tertiary Period, the isolation of South America, which had lasted some 40 to 50 million years, was ended by the reestablishment of the land connection between North and South America. Mammals of both regions then began to spread northward and southward, migrating from one continent to the other. Not all species migrated, however; nor were those that did equally successful in the new continent, whether North or South. The North American mammals, though, were able before long to occupy many of the niches previously occupied by the indigenous South American species, leading to the replacement and extinction of many of the South American mammals (Table 3-1).

The present-day composition of the South American mammalian faunas is, therefore, a result not only of the particular environmental conditions found there, but also of the evolutionary history of the area during the Tertiary Period.

Table 3-1 Characteristic Families of Land Mammals In South and North America*

	SOUTH AMERICA			NORTH AMERICA		
Epoch	Total	Indigenous	North American	Total	Indigenous	South American
Holocene	30	16	14	23	20	3
Pleistocene	36	23	13	34	26	8
Pliocene	25	24	1	27	26	1
Miocene	23	23	0	27	27	0

*The table shows the total isolation of North and South America during the middle Tertiary Period (Miocene Epoch), with each continent having its own characteristic mammalian families. During the Pliocene Epoch the land connection between the two continents was reestablished and there was considerable faunal interchange. North American mammals, however, replaced many of the indigenous South American forms. Consequently, about one-half of the present-day mammalian families in South American have North American origins, whereas less than 15 percent of the North American families have a South American ancestry.

From G. G. Simpson, 1953, Evolution and Geography, Ore. State System of Higher Educ., Condon Lectures, p. 27.

Thus, the South American mammalian fauna includes groups that were initially established there in the early Tertiary Period, such as the armadillo, tree sloth, and anteater; monkeys and rodents that were somehow able to cross the water barrier separating the two continents during the middle Tertiary Period; and late Tertiary arrivals from North America, such as field mice and various cats, as well as groups such as llamas and tapirs that have since become extinct in North America.

SUMMARY

Paleoecological interpretation depends on knowledge of the ecology of present-day animals and plants, which provides models and examples for analyzing fossils and ancient environments. The nature of the environment is all-important for organisms, for their survival and reproductive success depend of their ability to cope with the physical, chemical, and biological factors present in the environment.

The genetic constitution of organisms largely determines how well adapted to the environment an organism is, in terms of feeding, seeking shelter, finding mates, avoiding predators, and so on. Each habitat, or major life zone, will have a variety of species, each well adapted to individual niches within it. The morphology, or way in which an animal or plant is constructed, provides useful insight to the organisms' habitat (or address) and its niche (or how it makes its living).

Important inorganic factors in the environment which limit the distribution and abundance of animals and plants are temperature, dissolved gases like oxygen and carbon dioxide, sunlight, salinity, and substrate. Of these, the last—substrate—is especially important for paleoecology, because this is one factor that is recorded directly in rocks, for it is the sedimentary matrix that encloses the fossils. Moreover, the nature of the substrate correlates well with other environmental factors that are not directly recorded. Organic, or biological, factors include plant–herbivore relations, predator—prey interactions, and symbiosis. Analysis of feeding types and substrate relations is particularly helpful in understanding the interactions of organisms to their environment.

Not only do individual species interact with the environment, but aggregations, or communities, of species do, too. In most instances the species are mutually interdependent, forming a more or less closed ecosystem in which energy and matter are continually circulated.

The diversity of species within communities varies depending on whether the community is predominantly controlled by harsh or fluctuating physical–chemical factors—a physically controlled, low-diversity community—or by a variety of biological interactions—a biologically accommodated community of high diversity. Global diversity gradients that increase toward the equator as well as from onshore to offshore are interpreted as primarily shifting from physically controlled communities near the poles and nearshore, to biologically accommodated communities near the equator and offshore. History and barriers to dispersal are also important in determining the Earth's biogeography.

four

taphonomy

I look at the geological record as a history of the world imperfectly kept . . . of this history we possess the last volume alone . . . of this volume, only here and there a short chapter has been preserved . . . of each page, only here and there a few lines. (Charles Darwin, 1859)

Of all the arguments that could be presented against his theory of organic evolution by means of natural selection, Charles Darwin believed that the lack of abundant fossils intermediate between ancestral and descendant species was by far the strongest. In his *On the Origin of Species*, Darwin thus devoted a whole chapter to dealing with the "imperfection of the geological record" to which he attributed the absence of fossils that would clearly display the small intergradations between evolving species. As suggested by the opening quotation of this chapter, Darwin viewed the history of the Earth—and by direct implication, the history of life—as very imperfectly kept. We can modernize his metaphor of a book by imagining that a movie made of the last several billion years has been badly damaged, so that only a relatively small part of the whole is preserved. Only in the rarest instances would there by long enough strips of continuous film that had captured the evolution of one species to another—evolution that is slow by human time measurement, but very rapid geologically.

In this chapter, therefore, we discuss *taphonomy*, which is the study of the events and processes that occur to animals and plants from the time of their death until they are discovered as fossils. Recall from Chapter 1 that paleoecology is the study of the ecological factors that control the distribution and abundance of fossil organisms from birth to death. Taphonomy, on the other hand, is concerned with what happens to the organism from its death to its burial in sediment, to its alteration as the sediment becomes rock, and to its eventual discovery by a paleontologist. Although not a formal part of paleoecology, taphonomy, like environmental stratigraphy, is a corollary discipline. Obviously, any strict paleoecological interpretation depends on knowing how representative the fossil specimens are of the species to which they belong, and how representative the fossil assemblage is of the original, once-living community. We also want to be aware of any differential preservation that might have occurred which would change the relative proportions of individuals and species. Furthermore, we have to know if the fossils have been significantly transported and are thus a mixture of several different, distinct communities. In short, because we start with fossils and work back to the original living species, we must be able to estimate how "good" a sample the fossils are of the original life assemblage that we want to interpret.

The following example well illustrates the problem of how taphonomic sampling bias can influence paleontological interpretation. Several paleontologists over the last two decades have made careful estimates of the changing abundances of animals and plants throughout the almost 600 million years of Phanerozoic time. Graphs of marine animal groups, for instance, show a rapid increase in numbers in the early Paleozoic Era, a levelling off in the middle Paleozoic Era, and a sharp decline at the end of the Paleozoic Era. Thereafter,

FIG. 4-1 Graph showing changing abundance of species and sedimentary rock volume over the Phanerozoic Eon. The shape of both curves is rather similar, suggesting that the abundance of fossil species for a given geologic period is correlated to the amount of sedimentary rock for that interval of geologic time. In short, the more rocks there are, the more fossils are found. One explanation for this correlation is simply sampling bias of geologic processes in preserving sediments and their included fossils. Can you think of other possible reasons for this correlation? (After D. Raup, 1976.)

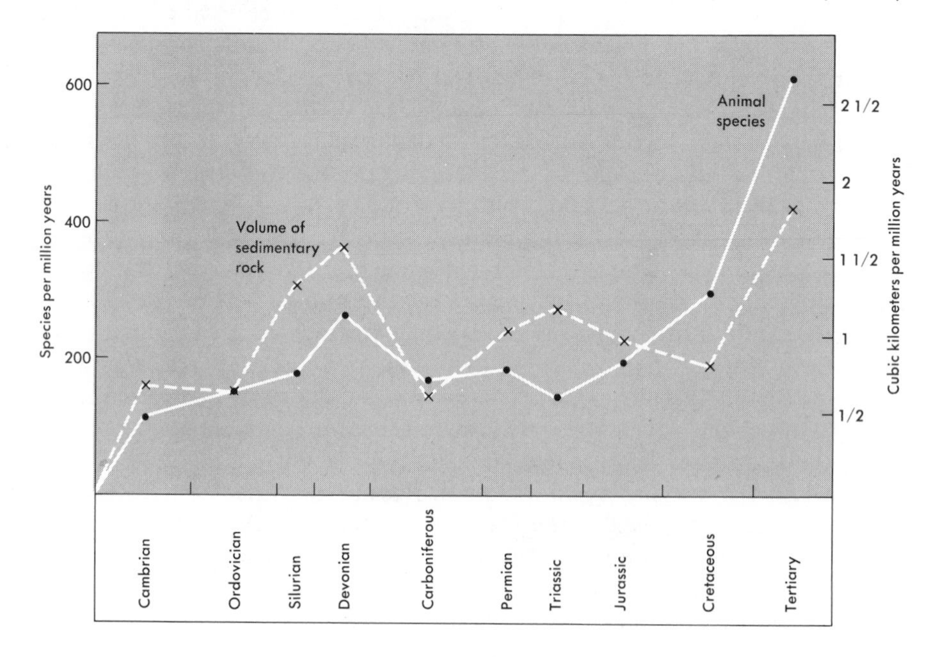

during the Mesozoic and Cenozoic Eras, there is a strong recovery and steady increase in diversity. These paleontologists suggest that the rise and fall and recovery of fossil life can be correlated to, among other things, expansion and contraction of shallow epicontinental seas; changing levels of oxygen and carbon dioxide; and fragmentation and reassembly of drifting continents. While these interpretations are often quite reasonable and intellectually appealing, some of their force was lost when another paleontologist showed that the changing abundances are mostly fortuitous, being correlated to the volume of sedimentary rocks preserved over the course of Phanerozoic time (Fig. 4-1).

SOURCES OF BIAS IN THE FOSSIL RECORD

As we discussed in Chapter 2 organisms that secrete hard parts have a much greater opportunity for fossilization than soft-bodied organisms. Hence, a fossil assemblage usually contains only the remains of the former, and is but a small sample of the total original community of species. For example, it has been estimated that of the some 10,000 species of animals and plants inhabiting a river-bank environment, only 10 to 15 are likely to be preserved as fossils in the river alluvium. For a coral-reef environment, perhaps only 50 to 75 species will be fossilized out of a possible 3,000 or more. Even fish, which have preservable hard parts (bones and teeth), are curiously absent from most reef sediments. Apparently the calcium phosphatic material is quickly oxidized or decomposed by microorganisms, and therefore rarely preserved. Yet, despite these losses, we can usually make important interpretations about the paleoecology of riverine and coral-reef environments from the fossils that are preserved.

Besides the preferential preservation of hard parts, there is another important source of bias in the fossilization process—the differential preservation of environments depending on whether an area is being eroded or is being buried with sediments. Table 4-1 summarizes the importance of hard parts and rapid burial in providing a good record of ancient life. For example, an assemblage of organisms living in an upland prairie is less likely to be buried and preserved in the rock record than an assemblage that lives on the continental shelf, because the upland-prairie environment is being more actively eroded than is the continental-shelf environment. As the land areas are being eroded away, the submarine areas are subsiding and their sediments are accumulating. Because of this selective preservation of habitats, it is not surprising, therefore, that the fossil record of marine clams is very good, because they secrete sturdy shells and live in environments where sediments are usually accumulating. Butterflies have virtually no fossil record, owing to the lack of preservable hard parts and their preference for habitats where erosion, not sedimentation, prevails. In general, therefore, the fossil record of marine calcareous algae and shelly invertebrates is many times more complete than that of animals and plants inhabiting land areas of high elevation, like prairies and plateaus.

Table 4-1 Factors Determining the Fossil Record

RAPID BURIAL	HARD PARTS YES	HARD PARTS NO
YES	Good clams	Poor jellyfish
NO	Poor birds	None butterflies

Besides lack of preservation owing to the absence of hard parts, organisms may be transported after death and buried in environments quite unlike those in which they lived. Many terrestrial organisms from a wide variety of habitats—woodland, grassland, flood plain—are found as fossils in sediments deposited in river channels, deltas, or lakes. In the oceans, too, skeletal remains of organisms living in shallow waters may be swept by storms or turbidity currents into much deeper parts of the ocean floor. Deep-sea cores, for example, often contain thin, graded, sandy layers of shelly debris of shallow water, benthic organisms that are interbedded with fine-grained, burrowed pelagic sediment from the deep sea. Obviously, turbidity currents occasionally move downslope, carrying shallow-water sediments and fossils well out into the ocean basins.

When an organism dies, oxidation and bacterial decomposition of the organic matrix in which its hard skeleton was secreted, or of the organic tissues which hold the hard parts together, result in the disarticulation of the individual pieces composing the skeleton. Thus do the many interlocked calcite crystals

FIG. 4-2 Experiments with five different kinds of vertebrate remains yield the equivalent hydraulic grain size for quartz spheres. For example, a horse molar behaves hydraulically as if it were a quartz grain 27 millimeters in diameter, whereas a dermal scute of a crocodile, perhaps one-third the diameter of the horse molar, behaves as if it were a quartz grain only 3 millimeters in diameter. The chief reason for these hydraulic differences is due to the varying densities of the assorted bones and teeth. (From Behrensmeyer, 1975.)

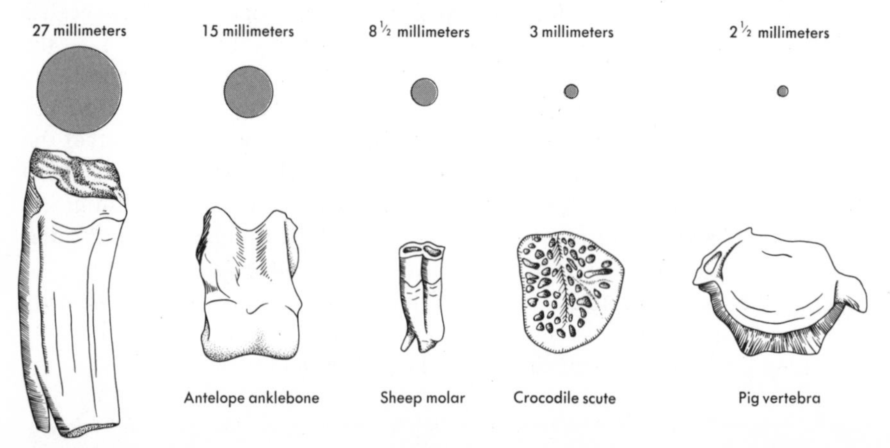

of a sea urchin become separated, or the two valves of a clam, or the various bones of a vertebrate. Because these disarticulated remains usually come in a wide range of sizes, shapes, and densities, some hard parts will be more easily transported than others by moving water, as illustrated in Fig. 4-2. This differential transportation of hard parts results in *hydraulic sorting* of the skeletal remains downcurrent. Such sorting is often noticeable on beaches where shells of animals living in the nearshore, subtidal environment may be preferentially accumulated by the surf and waves along the shoreline.

Table 4-2 Transportability of Sheep and Coyote Bones

GROUP I Removed immediately by saltation or suspension		GROUP II Removed gradually by traction		GROUP III Not removed at all
Ribs		Upper leg bone (femur)		Skull
Vertebrae		Lower leg bone (tibia)		Lower jaw
Breast bone		Upper arm bone (humerus)		
		Wrist/ankle bones		
		Pelvis		
		Lower arm bone (radius)		
	Shoulder blade		Lower jaw blade (ramus)	
	Finger/toe bones			
	Lower arm bone (ulna)			

(After Voorhies, 1969.)

The skeletal parts of terrestrial vertebrates are particularly susceptible to hydraulic sorting downstream in a river channel or along a sloping land surface. Experiments with coyote and sheep skeletons indicate which parts of the skeleton are easily transported, which not (Table 4-2). The skull and lower jaw of sheep and coyotes are rather massive and thus not easily transported, whereas the disarticulated ribs, vertebrae, and breast bone are moved readily owing to their less dense character. The experiments from which these data are derived give us some insight about how to interpret the ancient environments of fossil vertebrates. If a fossil assemblage includes bones from Group I as well as from Groups II and III, we would feel confident in concluding that the sedimentary rocks enclosing these fossils record not only the depositional environment but the original life environment as well. If, however, the vertebrate assemblage is almost completely composed of Group I bones, then we ought to suspect that the bones have been transported, perhaps into an environment quite unlike the original life environment of the animals.

Differential transportation seems most likely in high-energy environments; that is, where fast-moving currents can pick up and carry off skeletal remains.

Typical environments include areas of high local relief (on land and in the sea), along streams and rivers, and along beaches and tidal flats where there are strong waves and currents. Lower-energy environments, on the other hand, as those found in lakes and offshore, appear less likely to sort skeletal remains differentially. For example, in a number of different shallow marine environments, shelly remains that are medium sand size and coarser accumulate in the same environments where the living shelly invertebrates and calcareous algae are found (Table 4-3). Apparently, only skeletal debris that is finer-grained than medium sand—fine sand, silt, and clay-sized particles—is easily transported and deposited in other depositional environments.

Table 4-3 Living and Dead Shelly Invertebrates, Main Channel, Mugu Lagoon, Southern California

Species	A Live	B Dead
Sanguinolaria nuttallii	676	454
Cryptomya californica	204	294
**Dendraster excentricus*	42	54
Dipolodonta orbella	15	5
†*Olivella biplicata*	3	16
Chione californiensis	2	6
Spisula dolabriformis	1	2
†*Nassarius fossatus*	1	1
†*Lunatia lewisi*	1	1
†*Polinices reclusianus*	1	1

*Echinoid
†Snails
All the other species are bivalves.
(*From Warme, 1971.*)

There is a third major source of sampling bias in the fossil record that is part of *diagenesis*, which is the process by which loose, wet sediments become dense, hard rocks. After a sediment is laid down, it is usually buried by successive deposits as the area of sedimentation continues to subside. As burial continues, the weight of the overlying, accumulating sediment causes the materials below to undergo *compaction* and *consolidation*. Individual sedimentary grains are pushed and crowded together, thus reducing the initial pore space that existed between the grains. In the case of water-laid sediments, the interstitial fluids are slowly squeezed out as the porosity of the sediment diminishes—sometimes resulting in a volume decrease of 50 percent or more. The interstitial fluid is originally incorporated from the transporting medium (sea water or fresh water) during sedimentation, and is identical with it in terms of chemical composition, acidity, and oxidizing potential. With time, however, the interstitial water may radically change its chemical character. Much of this change is due to the solution of unstable minerals and the decomposition of included organic matter by micro-organisms.

As the coarser-grained sediments are compacted, the pore water moves upward through the sediments and may dissolve or precipitate mineral matter along the way. Thus, compaction of sediments with subsequent squeezing out of the interstitial fluids initiates the next step in the diagenetic process, namely, *grain cementation*. Individual sedimentary grains are welded together by mineral matter—usually silica or calcium carbonate—deposited by the interstitial fluids. Cementation may continue, especially if the sediments are later uplifted and located within the zone of freshwater precolation and saturation where large quantities of fluids may freely circulate.

Muddy sediments, on the other hand, after compaction, undergo *recrystallization*, which is likely to be more prevalent in these finer-grained sediments than grain-to-grain cementation. Although the kinetics of this process are not fully understood, it appears that as muddy sediments are consolidated, individual mineral grains are brought close together and, through a reorientation of their atoms and molecules, recrystallize into a dense and almost impermeable framework. This spontaneous recrystallization may be triggered by chemical equilibrium changes brought about with increasing pressures and temperatures as the sediments are buried deep within the upper part of the Earth's crust.

Now, we want to consider what happens to the organisms that are buried in the sediments that are undergoing diagenesis. Any soft-bodied remains that survived the effects of erosion, scavenging, and decomposition will be squeezed flat during sediment compaction. If the fluids circulating within the sediments contain oxygen, the oxygen itself as well as bacteria supported by the oxygen will destroy the remains. If, however, there is little or no circulating oxygen, the buried remains may well be preserved as thin carbon films. Such preservation of carbonized remains are typical of stagnant or low-oxygen depositional environments, such as deep lakes or deep marine basins. Hard parts that are composed of calcium phosphate, like vertebrate bones, teeth, and the shells of some invertebrates (refer back to Table 2-2), are also more likely to be preserved in sediments from stagnant depositional environments, where the absence of oxygen inhibits bacterial decomposition and oxidation of the remains.

Hard parts composed of calcium carbonate—the minerals aragonite or calcite—and opaline silica are susceptible to solution during diagenesis. Interstititial water will often dissolve these materials, sometimes leaving behind molds in the sediment (or rock) which encased them or else replacing the original shell material with casts of silica or calcite. Some skeletal mineralogies are much more susceptible to solution than others. For example, calcite rich in magnesium is about ten times more soluble than aragonite, which in turn is several times more soluble than calcite with little or no magnesium in the crystal structure. Obviously, then, an assemblage of potential fossils that includes aragonite (green calcareous algae, corals, some clams and snails), high-magnesium calcite (red calcareous algae, some forams, and some echinoderms), and low-magnesium calcite (oysters, barnacles, and other echinoderms) may experience differential solution, so that only the low-calcite fossils are ultimately preserved. Hence, in this instance, not only are the soft-bodied organisms not preserved, but even some of the hard-part secretors fail to survive diagenesis.

There is a final kind of bias in sampling the fossil record, and it originates in the mind of the paleontologist who is searching for fossils. Two examples will illustrate what we mean. For a long time, paleontologists looked for and could not find fossils in Precambrian rocks. What they were looking for were either the shells of early invertebrates that were presumably ancestral to Cambrian shelly invertebrates, or else the carbonized remains of their soft-bodied ancestors. When none of these were found, a number of hypotheses were formed to explain their absence: the rocks were too metamorphosed to preserve such fossils; the environments in which these organisms lived were not preserved; or, a long interval of erosion occurred during which rocks with the fossils were lost. In all cases, the search was predicated on the belief that the fossils would be of similar size and kind to those found in younger rocks in early Paleozoic time.

Then, in the 1950's, first by chance, later by intention, people began to find a fair abundance and variety of microscopic, single-celled primitive fossils in Precambrian rocks. Once it was realized that there were no macroscopic, early Paleozoic-like fossils in Precambrian time, because life had not evolved to that level of organization until just before the Cambrian Period, paleontologists became much more successful in their search.

Another example of "intellectual bias" in the fossil record comes from trace fossils, which are the indirect remains of organisms—burrows, tracks, and trails. While such fossils had been known for a long time, only a small proportion were truly recognized for what they were. Many such fossils were incorrectly attributed to algal impressions, for example. But again, once paleontologists understood the origin and biological and ecological significance of such structures, virtually every kind of sedimentary rock revealed a number of such traces. Part of the recognition problem was due to the necessity of looking for these fossils in a new way, namely, cutting and polishing rocks at right angles to the bedding so that these traces were more apparent. Also, many rocks containing traces lack body fossils; that is, shells, bones, or teeth. Because such rocks were thought to be "unfossiliferous," paleontologists didn't bother with them. One wonders what other fossils we may be ignoring simply because we lack the imagination to look for them or recognize them when we see them.

Given the sampling bias in the fossil record owing to taphonomic events and processes, what hope do we have in making reasonable sense out of that record? Can we expect to reconstruct ancient environments in any meaningful way if the original assemblage of animals and plants is so poorly and unevenly preserved? In the rest of this chapter we discuss two examples that demonstrate we can, indeed, come to some positive conclusions about ancient life and its environments in spite of the sampling bias inherent in the preservation of fossils and rocks.

AN OLIGOCENE OYSTER COMMUNITY

In order to test specifically just how much and what kind of information is lost during fossilization, a paleontologist examined an Oligocene oyster bank in North Carolina and compared the fossils found there with living oyster banks in the same region. The fossil oyster bank of Late Oligocene time is lens-shaped and measures two and one-half meters high and more than 40 meters long. The oysters initially settled in a two-meter deep, nearshore channel and built upward and laterally with the passage of time. Sandy sediments accumulated next to the bank and gradually filled the channel, covering most of the oyster deposit. The most abundant fossils in the oyster bank are large oysters, *Crassostrea gigantissima*, which are packed densely together, many of them in their original life position with the left valve encrusting the substrate (usually other oysters or shelly invertebrates). Associated fossils include foraminiferans, encrusting bryozoans, spirorbid worms that secrete calcareous tubes, snails, barnacles, and echinoid debris. Besides these body fossils of shelly invertebrates, there are also trace fossils that include borings made by bivalves, bryozoans, sponges, and worms, as well as internal partitions and blisters on the inside of the oysters themselves that record efforts by the oysters to seal off soft-bodied worms that were preying upon them.

Notice that not only do the body fossils tell us which organisms were originally present in the Oligocene oyster community, but the distinctive boring patterns and internal partitions and blisters provide additional evidence of other organisms living in the community. These latter organisms are not fossilized directly because either they were soft-bodied—as in the case of the various worms and sponges—or their aragonite shells were subsequently dissolved—aragonitic bivalves and snails. Such indirect evidence of the original members of the community is *redundant*, in that there is a duplication of information about an organism's initial presence, first by preservation of its hard parts or else by preservation of indirect evidence like traces or borings.

Having established the composition of the fossil oyster community, both by direct evidence of body fossils and by indirect, redundant evidence of soft-bodied and aragonitic forms, published reports on the faunal composition of modern oyster communities found along the North Carolina coast were examined. The total number of species ranges from around 50 to over 200, depending on the local salinity of the water, with more normal marine saline waters having the higher diversity. Of these species, 80 are quite common and seem to be most characteristic of present-day oyster communities there. These 80 species are listed in Table 4-4, and for each it is indicated whether it is soft-bodied or not, and the chemical nature of the hard parts. It is also indicated which species may leave behind redundant information. For example, all five species of sponges secrete microscopic siliceous spicules that reinforce the protoplasmic mass. Not only might these spicules be found as fossils, but three of the species also bore into shells so that their borings would provide redundant information about their presence. Recognition of such redundancy is important because the fine-grained, opaline silica of sponge spicules is quite soluble and, consequently, commonly not preserved. This is in fact the case in the Oligocene oyster community, and only the redundancy of sponge borings established their presence. Slightly more than half of the total common species have hard parts; and of these, more than one-third are capable of leaving redundant information.

Table 4-4 Fossilization Potential of a Modern Oyster Community

Taxa	Total Species	Soft-bodied	With Preservable Hard Parts				Possible Redundancy
			Ca	Ch	Si	Ph	
Porifera	5	—	—	—	5	—	3
Coelenterata	6	5	1	—	—	—	—
Platyhelminthes	1	1	—	—	—	—	—
Nemertea	2	2	—	—	—	—	—
Bryozoa							
Ectoprocta	7	4	3	—	—	—	—
Annelida							
Polychaeta	13	13	—	—	—	—	4
Mollusca							
Gastropoda	9	—	9	—	—	—	1
Pelecypoda	13	—	13	—	—	—	2
Arthropoda							
Crustacea	19	10	4	5	—	—	5
Arachnida (?)	1	1	—	—	—	—	—
Insecta	1	1	—	—	—	—	—
Chordata							
Tunicata	2	2	—	—	—	—	—
Vertebrata	1	—	—	—	—	1	1
Totals	80	39	30	5	5	1	15
Percentages of total community	100	49	38	6	6	1	19

Among the arthropods, only decapod crabs with relatively well-calcified and/or well-tanned exoskeletons have been included with the organisms with hard parts.
Ca = calcareous
Ch = chitinous
Si = siliceous
Ph = phosphatic

(*From Lawrence, 1968.*)

Next, we can tabulate the kind of calcareous skeletons these living species secreted; whether they were all or mostly calcite, all aragonite, or significant mixtures of the two mineralogies (Table 4-5). This is done to estimate the possible loss of information that would result if the calcareous hard parts were differentially dissolved. Notice that 30 percent of the 80 species secrete either calcite or aragonite in roughly equal proportions (11 species to 13 species), whereas eight percent secrete a skeleton which is a mixture of the two minerals.

Table 4-5 Effect of Aragonite Loss on Preservation of Modern Oyster Community

Taxa	Number of Species	A	C	A + C
Coelenterata	1	—	1	—
Bryozoa	3	—	3	—
Mollusca				
Gastropoda	9	9	—	—
Pelecypoda	13	4	3	6
Arthropoda				
Crustacea	4	—	4	—
Totals	30	13	11	6
Percentage of total community	38	16	14	8

A = completely aragonitic skeletons
C = entirely calcitic skeletons; calcite skeletons with very minor aragonite, as in crassostreids
A + C = appreciable aragonite plus calcite in skeletons
(*From Lawrence, 1968.*)

Finally, we can follow the three taphonomic stages that a modern community might pass through during its fossilization history, from the original community with all organisms present, to one where only hard parts and redundancy are preserved, to one where only calcitic remains and redundant information are recorded (Table 4-6). The state of preservability of each of these stages is then compared with what is found in the Oligocene oyster community (column D in Table 4-6). These data strongly indicate that the Oligocene community is quite similar in its composition to a modern oyster community if the latter were to be fossilized. In other words, even though the Oligocene fossil assemblage has fewer fossils, both in kinds and abundance, than the modern community, we can explain this loss of fossils by *predictable* taphonomic processes rather than attribute it to significant differences in original composition of the marine faunas. This example, therefore, illustrates the importance of taphonomic analysis in estimating what the original fossil community was like before its subsequent fossilization. As paleoecologists, we want to explain the variations in distribution and abundance of fossil organisms in terms of the original environment. But first we must have a good idea of how much those variations are due to differential preservation and the extent to which they are due to primary environmental causes.

Table 4-6 Taphonomic Stages in Preservation of Modern Oyster Community and Comparison with Oligocene Example

	TAPHONOMIC STAGES			OLIGOCENE COMMUNITY
	A →	B →	C	D
Phyla represented	9	7	7	4
Species present	80	45	18	16–18
Percentage of total community preserved	100	56	23	—
Percentage of information through redundant transmissions—preservation of nonbody parts	0	7	33 ↑ (compare)	~44 ↑ (compare)

Column A: original community; all organisms preserved
Column B: all hard parts and all redundant information preserved
Column C: aragonitic, mixed aragonitic-calcitic, chitinous, and siliceous skeletons lost; redundant information preserved
Column D: Oligocene community, for comparison with C
(*After Lawrence, 1968.*)

A PLIOCENE VERTEBRATE COMMUNITY

Another interesting test case of the reliability of the fossil record, given its bias for differential fossilization, comes from fossils and recent mammals of western Montana (Tables 4-7 and 4-8 and Fig. 4-3). First a census was made of the present-day mammalian assemblage living in and around a small river valley that lies between two mountain ranges. Making allowance for changes in the fauna due to the effects of twentieth-century civilization (hunting, trapping, farming, spraying of pesticides, and so on), there are in the area 49 species of mammals, including beaver, squirrels, bears, wolves, mice, rabbits, badgers, and elk. Of these 49 species, only 14 are likely to be preserved as fossils. The remaining 35 species will be absent or poorly represented in a future fossil assemblage, because they are very small, nonburrowing forms (various mice, for example) or have few representatives (bears and mink).

Next we can estimate the population density—high, medium, or low—and habitat preference—river terrace, flood plain, river bank, or conifer forest—for each of the 14 species. In this way we are able to predict the probability of finding any given species as a fossil in a specific habitat.

Having established the predicted frequency of these mammals in various river-valley and associated sediments we can then collect the mammalian fossils from Middle Pliocene valley sediments of the same area. After identifying the fossil mammals and estimating their density in the different Pliocene habitats as inferred from the sedimentary rock matrix, we find a striking similarity between the Pliocene mammal community and the recent one. Not only are the Pliocene and recent habitats analogous in kinds and numbers but the relative

Table 4-7 Niches and Population for Holocene Mammals*

Animal	Food Habits	Population Density (d)	HABITAT ZONES Terrace p	Terrace dp	Flood Plain p	Flood Plain dp	River Bank p	River Bank dp	Coniferous p	Coniferous dp
Pronghorn	Large grazer	2	9	18	7	14				
Bison	Large grazer	2	6	12	8	16			2	4
Rocky Mountain sheep	Large grazer	1	8	8	8	8				
Whitetail jack rabbit	Small grazer	3	8	24	6	18			2	6
Cottontail rabbit	Small grazer	3	5	15	6	18			5	15
Mule deer	Large browser	2	3	6	5	10			8	16
Whitetail deer	Large browser	1	3	3	8	8			5	5
Elk	Large browser	1	6	6	7	7			3	3
Moose	Large aquatic plant-eater	1			7	7	6	6	3	3
Muskrat	Small aquatic omnivore	2					16	32		
Beaver	Small aquatic bark-eater	1			2	2	12	12	2	2
Pocket gopher	Very small burrowing granivore	3	5	15	8	24			3	9
Ground squirrel	Very small semi-burrowing granivore	3	6	18	6	18			4	12
Ground squirrel	Very small semi-burrowing granivore	3	6	18	6	18			4	12
Total = 14 species		s		11		13		3		11
		t (dp)		133		174		50		91
		% t (dp)		29%		39%		11%		20%

d—total species population density weighted from 1 (low) to 3 (high)
p—species habitat-zone preference weighted from 0 (none) to 16 (total)
dp—species population density within a particular habitat zone
s—species representation, *i.e.*, number of species occurring in a particular habitat zone
t—habitat-zone population density relative to total population density
% t (dp)—percentage of habitat-zone population density relative to total population density

*The table shows the ecologic niches and population densities in river valley habitats for recent mammals likely to be fossilized. Species in italics are those forms in the Pliocene Epoch which have been replaced ecologically by different modern species. Note the close similarity between predicted values of %t(dp) for recent mammals and observed values from Pliocene fossils (Table 4-8).

From R. L. Konizeski, 1957, Geol. Soc. America Bull., *v. 68, pp. 131–150.*

Table 4-8 Niches and Population for Pliocene Mammals*

Animal	Food Habits	Population Density (d)	HABITAT ZONES Terrace		Flood Plain		River Bank		Coniferous	
			p	dp	p	dp	p	dp	p	dp
Pronghorn	Large grazer	2	9	18	7	14				
Horse		2	9	18	7	14				
Horse		2	9	18	7	24				
Camel		2	9	18	7	14				
Camel	Large browser-grazer	2	7	14	9	18				
Rabbit	Small grazer	3	8	24	6	18			2	6
Peccary	Large rooter	2			8	16	4	8	4	8
Peccary		2			8	16	2	4	6	12
Proboscidean	Very large aquatic plant-eater	1			6	6	10	10		
Rhinoceros	Large aquatic plant-eater	1			6	6	10	10		
Beaver	Small bark-eater	1			1	1	14	14	1	1
Ground squirrel	Very small semifossorial granivore	3	8	24	2	6			6	18
Ground squirrel		3	8	24	2	6			6	18
Total = 13 species,		s		8		13		5		6
111 specimens		t (dp)		158		149		46		63
		% t (dp)		38%		36%		11%		15%

*The table shows the ecologic niches and population densities in river valley habitats for Middle Pliocene fossil mammals. All footnotes to Table 4-7 apply to this table.

abundance of species in terms of adaptive types is essentially similar. The ecologic niches filled during Pliocene time by horses, camels, rhinoceroses, and proboscideans are today filled by ecologically similar forms such as deer, elk, and moose. The most significant conclusion of this study is that a fossil community can be reconstructed from some, not necessarily all, representatives of the fauna and flora. This conclusion reaffirms what many paleontologists have intuitively felt to be true all along, namely that despite the loss of many members of an original assemblage of organisms, the few remaining fossilized species are usually adequate for environmental reconstruction.

This example has demonstrated that the ecologic roles that Pliocene and recent mammals played in river valley environments has remained essentially the same. The "actors" filling some of the roles, however, have in some instances changed, although the roles remained the same. Thus, the ecologic role of river-bank-dwelling aquatic-plant feeder was filled by proboscideans and rhinoceroses in the Pliocene Epoch; today that same role is filled by a different taxonomic group, a species of moose.

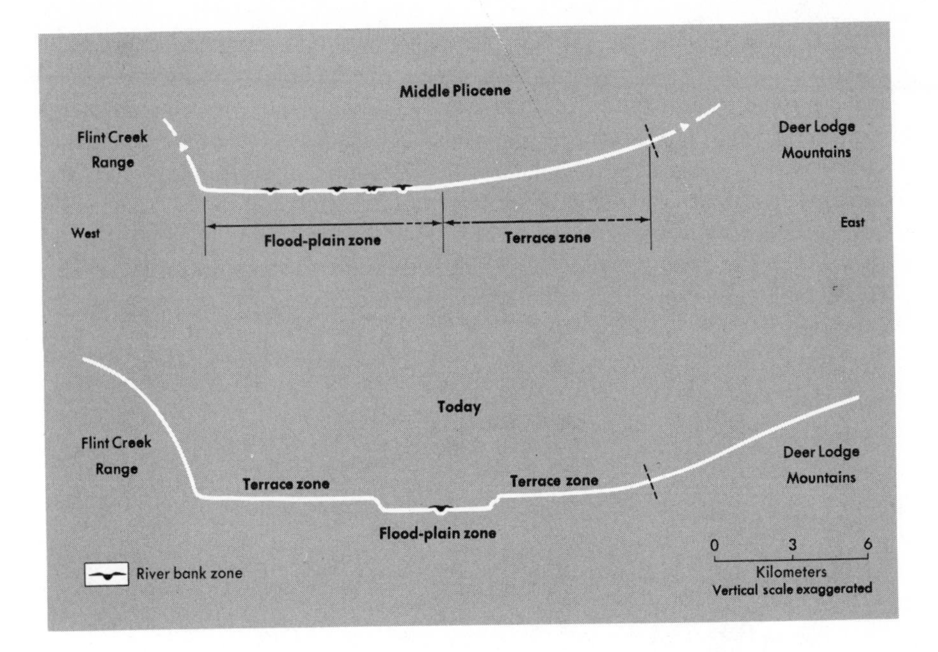

FIG. 4-3 Inferred distribution of Middle Pliocene habitats (top) and generalized distribution of the same habitats today (bottom). (From Konizeski, 1957.)

SUMMARY

As paleoecologists, we want to understand the interactions of fossil animals with their ancient environments. Clearly, to do this we have to have a fair idea of what the original community of organisms was like. But owing to the lack of preservation of all the original members of the community, we only see preserved as fossils a small sample of the original assemblage. Therefore, an essential part of paleoecological analysis requires that we consider the various taphonomic processes and events that may have produced the fossils we find from the original living animals and plants. In short, we have to take account of the inherent bias in the fossilization of the former organisms.

Fossilization introduces several kinds of bias: organisms without skeletal hard parts are rarely preserved directly, although they may leave redundant information in the form of tracks, trails, burrows, borings, or pathologic markings that indicates their original presence. Hard parts themselves may be sorted and transported by currents so that fossils from different life environments are mixed together in the final depositional environment in which they are buried. Hard parts may also be preferentially preserved owing to differing resistance to oxidation and solution of common fossil materials like calcite, aragonite, opaline silica, and chitin.

Despite the various kinds of sampling bias that taphonomic processes introduce in the fossil record, they needn't necessarily prevent our understanding and interpretation of that record, particularly if we are alert to the problem. In fact, taphonomy helps explain why the fossil record is the way it is, and why it is not merely a haphazard accumulation of shells and bones. The taphonomy of an Oligocene oyster community and a Pliocene river-valley community not only explains how the original communities ended up the way they did in the fossil record, but also how those fossil samples can be reliably interpreted even though not all the original members of the communities are preserved.

five

environmental analysis

Surprisingly detailed interpretations of past environments are possible, and the fossils themselves acquire much added significance whenever they are studied in the full context of their geologic setting. (Norman D. Newell, 1959)

In previous chapters we have reviewed some of the relationships among organisms, sediments, and the local depositional environment. But knowledge of these relationships alone will not enable us to reconstruct an *ancient* environment. We also need to develop some insight into just how we go about collecting and examining fossils and rocks so that we can arrive at meaningful conclusions and significant interpretations about ancient life and its environmental setting.

In this chapter, therefore, we will consider some of the assumptions, procedures, viewpoints, and limitations that form the background for paleoecological investigation and interpretation—to ask, in other words, in what sort of general intellectual framework are paleoecological studies pursued? In earlier chapters we have discussed paleoecological tactics; here, we will discuss paleoecological strategy.

THE PRESENT IS A KEY TO THE PAST

Up until about the beginning of the nineteenth century it was generally believed that fossils within rocks were sports of nature, works of the devil, or the remains of animals and plants buried by the Biblical Deluge. Among the more educated, however, fossils were indeed recognized as the petrified remains of once-living organisms. But it was commonly assumed that these organisms had been killed by successive and sudden geological "revolutions" during which the seas inundated the continents, oceans dried up, and the Earth's surface underwent great upheavals. Such beliefs were summed up by paleontologist Georges Cuvier (1769–1832) in his theory of *Catastrophism.*

At about the same time several Scottish geologists, including James Hutton, James Hall, and John Playfair, were in the process of formulating an opposite point of view, namely that the Earth undergoes constant change that takes place inexorably over great spans of time. The death knell of the catastrophic school of geology was sounded by the publication in the 1830's of the *Principles of Geology* by Charles Lyell, another British geologist. Lyell's particular contribution to the development of the young science of geology is best summed up in the subtitle to this work: ". . . an attempt to explain the former changes of the Earth's surface by reference to causes now in operation." Thus Catastrophism, which emphasized sudden and violent changes intermittently during Earth history, was replaced by *Uniformitarianism*, which emphasized slow, regular changes by processes similar to those observable today.

Since Lyell's time, geologists have realized the value of studying present-day Earth processes in order to understand ancient geologic phenomena, as expressed in the phrase, "the present is a key to the past." Although there are limitations to such an approach (for there may be some past geologic events that were truly unique and have no modern counterpart), paleoecology has depended strongly on this uniformitarian viewpoint. In fact, a logical outgrowth of the uniformitarian approach is the current reliance by paleoecologists on present-day organism–environment relationships to support inferences about the geologic past.

Paleoecology, as an historical science, thus has two separate, but interrelated aspects. On the one hand, there are the "past configurations" of fossil animals and plants in a wide variety of sedimentary rocks spanning hundreds of millions of years and found around the globe. On the other hand, if we want to understand how these particular past configurations came to be — and therefore go beyond mere description—we must have some body of theoretical concepts and empirical observations to guide us. Our only recourse, obviously, is the body of knowledge explaining how present-day animals, plants, environments, and geography are all interrelated. In the spirit, if not the letter, of Uniformitarianism, we believe that the principles and processes governing the distribution and abundance of life today can be invoked to explain life's past configurations. (And it is in precisely this spirit that the chapters preceding this one covered the topics they did.) This methodology of studying the recent in order to gain insight about the past is often referred to as *actualism*, to unburden it from some of the philosophical trappings of strict nineteenth-century Uniformitarianism.

It has become common practice now to cite a modern actualistic analogue when describing and interpreting an ancient environment. Many points of similarity between the recent and the ancient examples are sought, in the belief that the stronger the apparent similarity, the greater the probability that we are dealing with truly comparable environments, despite the often great separation in time and space. Even though the organisms in these past environments may be long extinct, we take it as a working hypothesis, at least, that the past depositional environment has some modern, extant representative, and that the present-day organisms stand in the same approximate relation to the environment that the extinct fossils do.

Table 5-1 provides such a comparison between a 400 million year old Devonian limestone of upstate New York with modern tidal-flat and shallow-water environments of South Florida and the Great Bahama Bank. In terms of sediment composition and texture as well as sedimentary structures, the depositional environments seem quite comparable. The organisms are not comparable for the most part, however, owing to the evolution of life during the long time separating the ancient example and its modern counterpart.

Table 5-1 An Ancient Environment and Its Modern Analogue*

Manlius Formation (Devonian, New York State)	Recent Analogue (South Florida and Great Bahama Bank)
1. Algal-laminated structures	Intertidal and shallow subtidal level in Florida Keys and Andros Island, Bahamas
2. Laminated, calcareous mud with pellets; often mudcracked.	Supratidal areas in Florida Keys, islands in Florida Bay, and Andros Island
3. Interbedded shell sands and calcareous muds; some dried mud-pebble conglomerates	Intertidal zone in Florida Keys and islands in Florida Bay
4. Scattered calcareous oölite sand	Just below intertidal zone, margins of Great Bahama Bank

After L. Laporte, 1967, Am. Assoc. Petroleum Geologist's Bull., *v. 51, p. 90.*

Having hypothesized the ancient depositional environment, we must next interpret the ecology (and/or taphonomy) of the fossils found there. Here the paleoecologist is on shakier ground, because while we can invoke principles and concepts from what we know of living organisms we cannot be totally confident that they apply to fossils, which are almost always extinct.

One particularly important point in paleoecological analysis is the realization that multiple lines of independent evidence must be sought and developed before any final conclusions are drawn about the nature of an ancient depositional environment. Because of the inductive and inferential nature of paleoecology, we can never definitely prove the truth of our assertions about past environments and communities of organisms. The only criteria we have for believing the validity of our paleoecological interpretations are, first the internal consistency of multiple sets of independent data which lead to the same final conclusions, and, second, the geologic and biologic sense our interpretation makes when compared to present-day environments and organisms.

Although it is the essence of the scientific method to formulate simple hypotheses to explain natural phenomena, we cannot simply perform a single set of critical observations and expect that all will become clear. However astute the observer and obvious the phenomena, geologic situations unfortunately have a tendency to be the result of many interacting variables, many of which are no longer directly recordable. Because sedimentary environments are inherently complex and multivariable, attempts to explain the abundance and distribution of organisms within a given environment by reference to any one single factor may be misleading and a serious oversimplification. For example, until recently some paleoecologists have attributed virtually all variation in the composition of fossil marine assemblages of similar age to differences in the original water depth of the ancient environment. Thus, whenever different fossils were encountered in a stratigraphic sequence of essentially the same age in a particular area, these differences were attributed to local changes in sea level. This explanation, in turn, required multiple and frequent sea-level fluctuations, an hypothesis that, for other paleoecologists, strained credulity.

Although it is true that the water depth of present seas sometimes correlates with changing marine communities, there are other, more direct, ecologic factors that limit marine organisms. For example, water temperature, oxygen and nutrient content, light, and substrate are usually more significant than water depth. The reason that depth variation is sometimes correlated with community variation is that these other, more direct ecologic factors are likely to vary with depth (refer back to Fig 3-11). Thus, substrates tend to be finer-grained, and light intensity decreases with increasing depth in the oceans. Yet many organic communities vary dramatically without any significant depth changes; the depth-control theory proves useless in explaining these variations. Such instances are the result of differences in critical ecologic factors that are, in these specific cases, independent of depth variation. For example, the marine communities of the Great Bahama Bank vary from area to area, yet these communities may be found in virtually all water depths that occur on the Bank proper. The ecologic factors determining the Bahamian communities are salinity, water turbulence and circulation, and substrate (see Fig. 3-9).

Once we have interpreted one environmental complex of sediments and organisms in terms of specific ecologic factors, we must resist the temptation to explain all similar complexes in the same way, since different environmental processes and controls may give roughly similar results. For example, red beds, or black shales, or reef limestones, each with characteristic lithologic and biologic attributes, may not necessarily form in always the same set of circumstances. It may well be that eventually paleoecologists will provide relatively simple, unifying theories for many different ancient environments. But until that time, the results of individual paleoecological studies must be cautiously applied to other situations, however similar they may at first appear.

Possible oversimplification, therefore, should make us wary of either placing too much value on a single environmental factor in a paleoecological interpretation, or assigning one set of paleoecological conclusions, however valid, to a second paleoecological problem merely because the two share some similarity of fossil composition and rock type.

SEDIMENTARY FACIES DEFINITION AND RECOGNITION

As we noted in Chapter 1 and 2 environmental stratigraphy is an important corollary to paleoecological analysis. Because of the intimate association of fossils in layered sedimentary rocks, certain stratigraphic preliminaries are required before the paleoecology of the fossils can be determined. These preliminary steps in environmental stratigraphy are the definition of the major facies within the rock whose fossils we wish to interpret paleoecologically, and the recognition of the important environmental parameters associated with each facies. Usually such facies definition and recognition require four separate steps (Fig. 5-1).

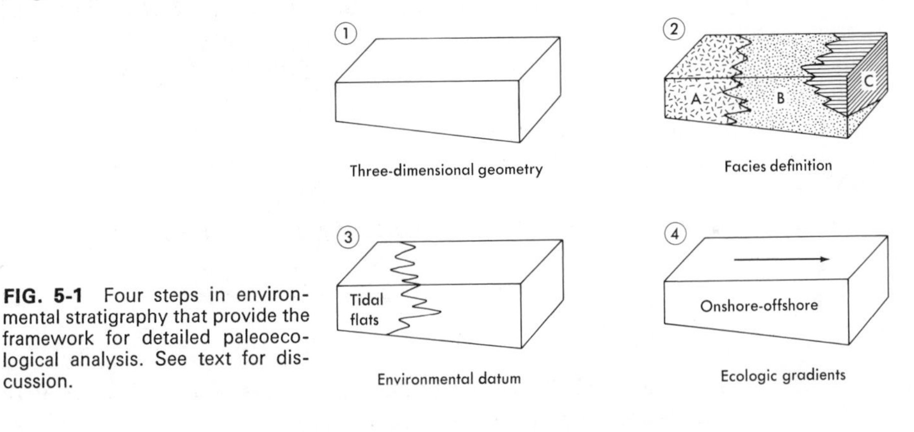

FIG. 5-1 Four steps in environmental stratigraphy that provide the framework for detailed paleoecological analysis. See text for discussion.

First, the three-dimensional geometry of the fossiliferous stratigraphic unit has to be defined. We need to know its areal extent, thickness, and its upper and lower boundaries. Next, we must define and delineate the lithofacies and biofacies in terms of the significant rock and fossil properties of the stratigraphic unit (refer back to Chapter 2). Properties of such lithofacies may include sediment composition, grain size, fabric or grain arrangement, color, bedding, and primary structures like cross-stratification, ripple marks, scour surfaces, and mud cracks. Biofacies characteristics include kinds of fossils—both body fossils and trace fossils—and their abundance; whether the fossils are whole or disarticulated; fresh or worn; reworked or in living position; evenly distributed or found in clumps or clusters; and so on. These lithologic and paleontologic variations within a stratigraphic unit presumably record original differences in the depositional environment where the particular sediments and fossils accumulated.

Third, from among the facies we have just defined, we look for one which has a modern analogue or counterpart that is reasonably well understood in terms of the environmental parameters responsible for its formation. This facies then becomes an *environmental datum* to which we can refer the other related, but perhaps less well-understood facies. Such a datum might be sun-dried mud cracks, coral reefs, freshwater calcareous algae, oolites, or evaporite salt deposits; any of these would pinpoint somewhere within the total paleoecologic complex the existence of certain environmental conditions. Rocks and fossils associated directly or indirectly with this datum could then be related ecologically to it, if only by the elimination of other, incompatible environmental possibilities. For instance, the finding of abundant and well-preserved reef corals in a limestone would immediately rule out a deep-water origin for the rock because reef corals require shallow, well-lit, agitated, and warm waters. The paleoecology of the other organisms that occur with the corals would thereby also be elucidated.

Having established an environmental datum for one facies, we can proceed to the final step in our environmental stratigraphic analysis, which is the prediction of ecologic gradients across the strike of the facies. Such gradients include nonmarine to marine; onshore to offshore; shelf to basin; closed or restricted circulation to more open, marine circulation; high turbulence to low turbulence; fine-grained to coarse-grained substrate; lowland to upland; more vegetation to less; and so on. We can choose among these various postulated ecologic gradients by seeing if the other laterally equivalent facies have attributes consistent with such a gradient. In other words, by looking for evidence of such gradients in these other facies we can come to some firmer conclusion about their depositional environments.

Returning to the example in Fig. 5-1, Facies A might exhibit such features as algal stromatolites, U-shaped burrows, mud cracks, scour surfaces with pebble-conglomerates, herring-bone cross-stratification, and other features found in modern tidal flats (refer back to Fig. 2-23). Knowing the origin of this facies, we can predict the kinds of ecological gradients from it into laterally adjacent facies. Thus, we would expect a tidal-flat facies to pass laterally into either a more marine, subtidal facies or a more terrestrial, nonmarine facies, depending, of course, on whether we are moving offshore or onshore from the environmental datum. If, say, an abundant and diverse marine facies were in Facies B, then an onshore to offshore ecologic gradient would be correct, as would be shallow to deep, or variable to normal marine salinity gradients.

DETERMINING ECOLOGIC GRADIENTS

Environmental analysis, like any other scientific analysis, is greatly enhanced if we can make our observations in the context of one or more working hypotheses about how the ancient environment might, in fact, be depicted. Formulating and testing such hypotheses will focus our attention on specific details that might otherwise be overlooked, as well as ensuring that we relate individual observations to some overall working model for the environment. Consequently, we will be immediately aware of inconsistent or contradictory observations, or of the absence of crucial missing pieces of the puzzle. Thus, contrary to popular image, scientists do not start out with a "blank mind" and collect data, and then see how they fit together. Even deciding what data to collect requires that we have at least some hunches—weak hypotheses—about the possible solutions to our problem. Of course, we mustn't be slaves to our original working hypotheses. On the contrary, we must expect that we will modify, or recast entirely, our initial hypotheses. As the late American humorist Robert Benchley once said, he knew he could build a bridge if someone could get him started. One way to get started in paleoecological bridge-building is to have a few working hypotheses ahead of time about the ancient environment we want to reconstruct and interpret.

One useful way of generating hypotheses about the environment we are studying is to attempt to predict the ecologic gradients that might have existed based on the environmental stratigraphy of the rocks in question. In this section, therefore, we will consider several kinds of ecologic gradients that we might expect to find.

Carbonate Epeiric Seas

It is possible to hypothesize the kinds of sedimentary deposits that would be laid down within an epeiric, or epicontinental, sea into which little or no erosional debris had been deposited. Such broad, shallow seas covered large areas of the continents throughout many intervals of geologic time, and thus provide us with some of the richest fossil horizons within the marine stratigraphic record. Such seas would have depositional slopes of very low angle, on the order of 20 centimeters or less per kilometer (one foot per mile). Such low slopes are quite unlike today's, when, during a period of relatively great continentality, the sea floors along continental margins have steeper depositional slopes of several meters or more per kilometer. It is then argued that, given such widespread, relatively shallow seas with low depositional slopes, a necessary consequence would be sharp gradients in water circulation and agitation across these epeiric seas (Fig. 5-2).

Different marine environments would soon develop approximately parallel to the ancient shore line that reflected an equilibrium between the local physical and chemical conditions and the associated sediments—in this case, carbonates.

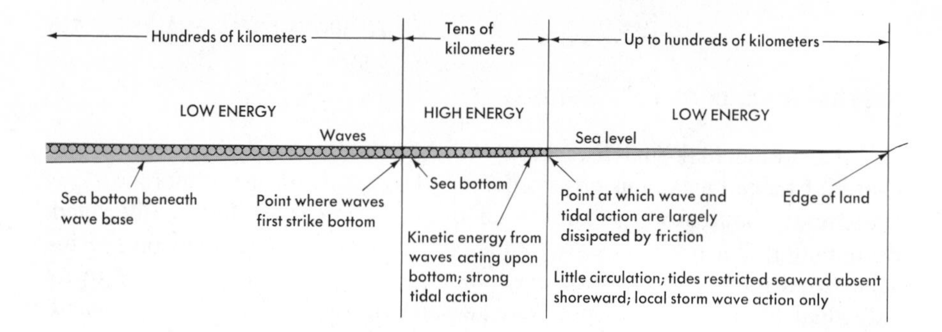

FIG. 5-2 Energy zones across a broad, shallow, epeiric sea. The sea floor slopes very gently—less than 20 centimeters per kilometer—and is divided into three environmental belts depending upon local hydraulic energy. The belts are offshore, below normal wave and tidal current action (X); onshore, where waves and tidal currents touch bottom (Y); and nearshore, landward of high energy zone, where water circulation and agitation are restricted (Z). Differences in water energy and mixing with more open ocean create, in turn, other important ecologic gradients in temperature, salinity, substrate, nutrients, turbidity, and so on. Compare with Figs. 3-9 and 5-3. (After Irwin, 1965.)

—and fossil organisms. Should sea level be rising or falling, these environmental belts with their individual lithofacies and biofacies would migrate accordingly, forming a vertical stratigraphy and locally recording the marine transgression or regression. (Recall Walther's Law discussed in Chapter 2.)

As shown in Fig. 5-3, these facies are chiefly differentiated into three major belts: an offshore area where the sea bottom lies below the zone of normal wave and current action; a second belt, more onshore, where waves and currents touch bottom and are dissipated; and a third, more nearshore belt, where wave and current action are virtually absent—except for occasional storms—even though the water depths here are very shallow, several meters or less. The individual widths of these facies belts depend, of course, upon the amount of slope on the sea floor and the magnitude of wave and current action. But even the most conservative estimates for slope and water energy indicate that these belts must have been tens to hundreds of kilometers wide.

FIG. 5-3 Carbonate facies in an epeiric sea that result from energy and related ecologic gradients. *Facies A:* Burrowed carbonate muds with diverse abundant marine fossils. *Facies B:* Oolites and skeletal sands with good cross-stratification; organic reefs might also be found here. *Facies C:* Burrowed carbonate muds; some pellets; fossils are less diverse but abundant. *Facies D:* Carbonate tidal-flat deposits; fossils are sparser still; mud cracks, scour-and-fill structures. *Facies E:* Evaporite deposits with few or no fossils, if arid environment where evaporation rates were high. Note how facies are *not* parallel to time lines. What would resultant vertical stratigraphy look like if sea level were rising? If it were falling? (After Irwin, 1965.)

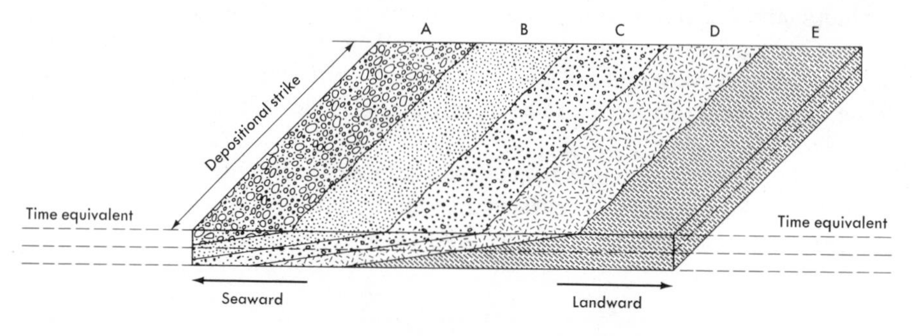

Within these facies belts, can we predict some of the other ecologic gradients that were most likely present? In the offshore, low-energy environment, we might expect marine waters of more normal marine temperature, salinity, and nutrients, owing to their connection with the more open, deeper, marine waters of the oceans. Here, sediments would tend to be finer-grained carbonate muds with an abundant and rich marine biota. The sediments would also likely be burrowed by the infauna, and only rarely reworked by waves and currents (storms?). Thus, burrow mottling of the sediments would predominate over primary structures like cross-stratification and ripple marks.

Further onshore, where the waves and currents touch bottom, we would expect coarser-grained carbonates, perhaps oolites and skeletal sands, with a more limited biota specially adapted to high water energy. (Can you think of some?) The innermost facies belt would again be fine-grained carbonate muds, with a more restricted biota, owing to more variability in temperature, salinity, nutrients, and so on. Depending on climate, we might expect to find evaporite beds nearshore; or, depending on degree of subaerial exposure, mud cracks and scour-and-fill structures (refer back to Fig. 2-25).

The model presented here is developed for epeiric seas where little or no clastic material enter. That is, the surrounding lands are of low relief and generate little erosional debris to the sea. We could develop a parallel model for an epeiric sea where, on the contrary, the surrounding lands stand high and contribute large amounts of clastics to the epeiric sea. (Could you predict the resulting ecologic gradients in terms of salinity, nutrients, substrates, water energy, and so on? How would the organisms vary in kind, number, and preservation along these gradients?)

Marine Trophic Analysis

Another conceptual model that is useful in predicting and recognizing important ecologic gradients in the fossil record is *trophic analysis*, which considers the variation of food resources on, in, and just above the sea floor, and how different bottom-dwelling invertebrates utilize them. That is, if we understand what food exists where on the sea floor, and who eats it, we can then apply this understanding to observed assemblages of fossil marine invertebrates.

Food in the sea, of course, comes in many different sizes (bacteria to large fish and whales), kinds (organic molecules and microorganisms to plants and animals), and conditions (live or dead, whole or piecemeal). Despite this diversity in food, or trophic, resources, however, most benthic marine invertebrates feed on the smaller particles that are dead or alive, animal or plant or microorganism. Given the low specific gravity of organic matter and its particulate nature, the distribution of food for benthic invertebrates varies with sediment grain size and water energy (refer back to Fig. 3-8).

In waters that are regularly agitated by waves and currents, there is abundant suspended food in the water column that supports suspension, or

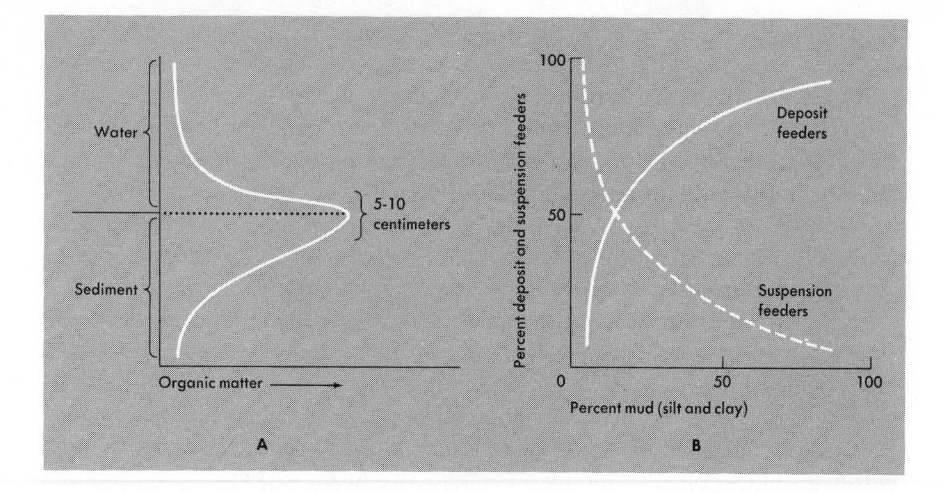

FIG. 5-4 (A) Concentration of organic matter—"food"—within five to ten centimeters of interface between sea floor and overlying water column. (B) Variations in suspension feeders and deposit feeders with changing mud content of the substrate. Mud-rich sediments are also food-rich, because the particulate organic matter in the water column settles out with the silt and clay. By contrast, mud-free sediments mean that the fine fraction of silt and clay is in suspension, as is the particulate organic matter; here suspension feeders are more common, filtering out the food from the overlying water column. (After Walker and Bambach, 1974; Purdy, 1964.)

filter, feeders. In quieter waters, the organic matter settles out and provides food for selective and nonselective deposit feeders on and within the sediment (recall our discussion of feeding types in Chapter 3). Two gradients in food resources are thus found in marine sediments: one that varies laterally with different water energy and concomitant changes in sediment grain size, and another that varies vertically from below the sea floor to the overlying water column. (Fig. 5-4).

Marine invertebrates utilize the variety of food resources in several particular ways as shown in Fig. 5-5. Some deposit feeders burrow deeply and feed on the included organic matter in sediments, while others feed on the same food in more shallow burrows. Suspension feeders, too, subdivide the food resources in the water column by filtering low or high off the substrate. Still others feed on algal material on the substrate (grazers) or eat somewhat larger particles of organic detritus found there (scavengers, or selective deposit feeders). Mobile predatory carnivores capture some of the foregoing animals as live prey. Notice that we can classify organisms by feeding type without particular reference to the taxonomic position of the organisms. Thus, organisms that differ taxonomically (like snails, bivalves, and trilobites) may use the same trophic resource of organic detritus on the sediment surface. Conversely, organisms of rather similar taxonomic position like echinoderms, may have very different feeding habits among themselves (crinoids, echinoids, and starfish).

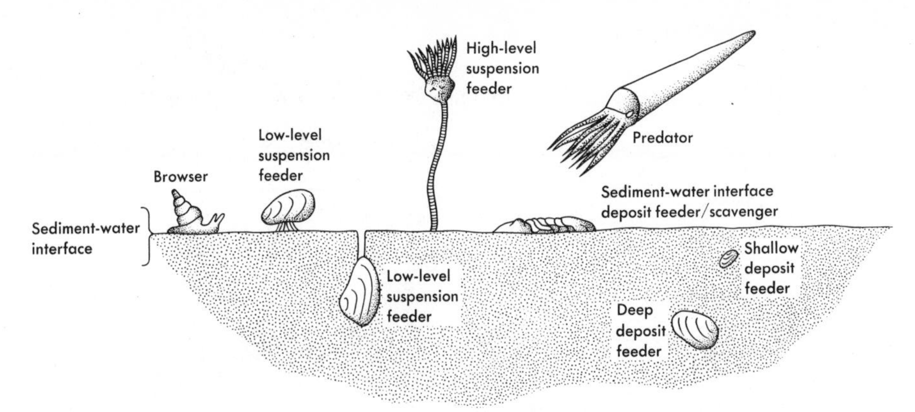

FIG. 5-5 Major feeding types among bottom-dwelling marine invertebrates. Notice how the variations in the kind and location of food resources are exploited by these feeding types. (After K. Walker and R. Bambach, 1974.)

Taxonomic Gradients

A consistent theme throughout our discussion of environments is the inherent lateral variation in ecologic parameters as we move across a geographic area. As just mentioned, we observe differences in marine food resources across varying substrates with differing water energy. Or we might see variations in salinity, temperature, oxygen content, and the other physical–chemical factors discussed in Chapter 3. Paralleling these ecologic gradients is the changing composition of the associated organisms. If we can monitor systematic changes in the general character of the kinds of taxa seen in the fossil record, we may be able to relate them to some appropriate ancient environmental gradient. An obvious example might be the increase of marine taxa and concomitant decrease

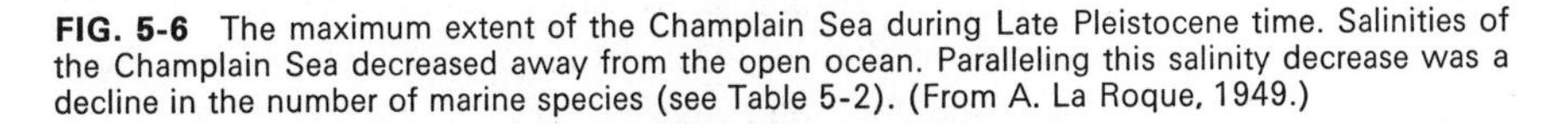

FIG. 5-6 The maximum extent of the Champlain Sea during Late Pleistocene time. Salinities of the Champlain Sea decreased away from the open ocean. Paralleling this salinity decrease was a decline in the number of marine species (see Table 5-2). (From A. La Roque, 1949.)

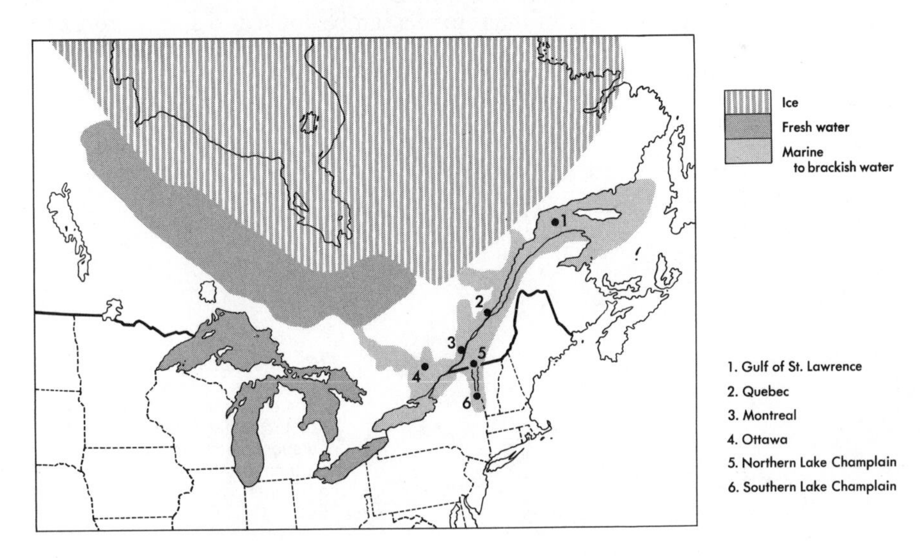

Table 5-2 Decrease in Diversity of Marine Species Going Away from Open Ocean

Number of Marine Species	LOCALITIES 1	2	3	4	5	6
Foraminifera	18	13	15	2	—	—
Sponges	1	—	2	1	—	—
Echinoderms	2	1	5	1	—	—
Bryozoans	26	5	2	1	—	—
Brachiopods	3	1	1	—	—	—
Clams	28	12	18	10	5	1
Snails	41	12	40	7	—	—
Annelid worms	11	1	2	2	—	—
Crustaceans	4	3	2	1	1	—
Total	134	48	87	25	6	1

From W. Goldring, 1922, N.Y. State Mus. and Sci. Bull 239, p. 164.

of nonmarine forms as we move across the strike of laterally equivalent facies, clearly indicating changes from a terrestrial environment to a marine one. Or, perhaps, the mere increase in diversity of marine taxa will suggest an increasing marine gradient (Fig. 5-6, page 109, and Table 5-2).

The proportion of planktonic to benthonic foraminiferans in modern seas provides a method for determining water-depth gradients in ancient rocks, in a rather general way at least. Present day planktonic forams increase sharply offshore, reaching a maximum on continental slopes and deep-water basins. Benthic forams, on the other hand, predominate in continental-shelf sediments and nearshore, with calcareous benthics more abundant on the shelves, and arenaceous forams—those that make their tests from sand grains—in nearshore brackish environments like marshes and lagoons (Fig. 5-7). These gradients in the taxonomic composition of microfossils can be applied to ancient rocks to

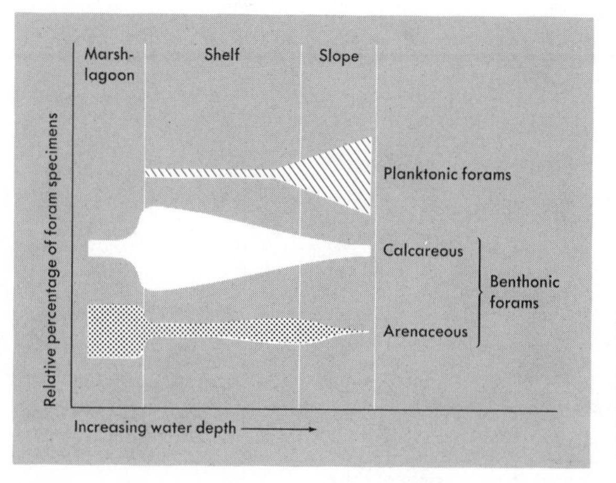

FIG. 5-7 Changing proportions of planktonic, calcareous benthonic and arenaceous benthonic forams with increasing water depth from nearshore environments across the continental shelf and down the continental slope. (From Eicher, 1969.)

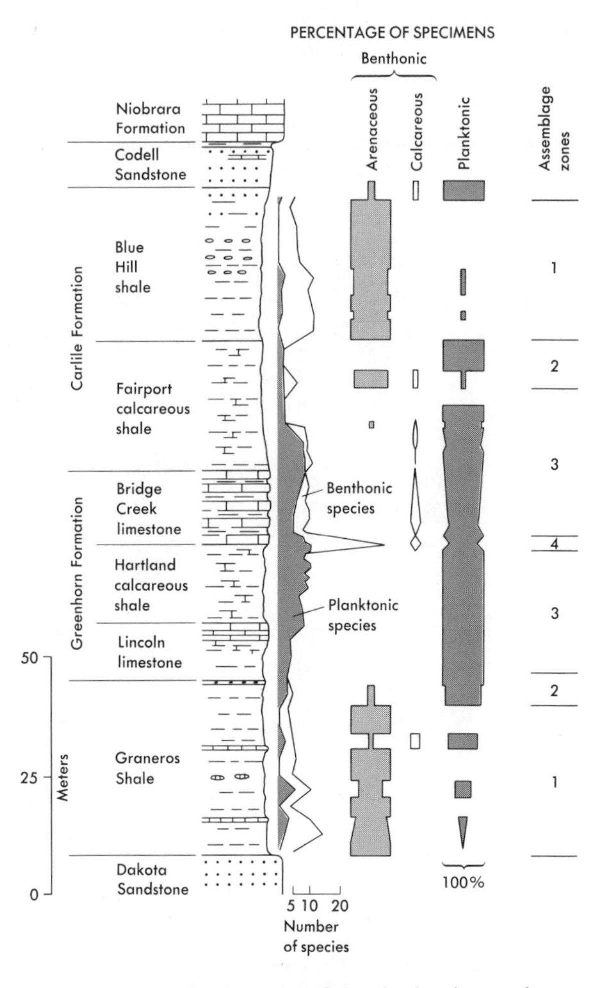

FIG. 5-8 Upper Cretaceous section in eastern Colorado showing varying proportions of planktonic, calcareous benthonic, and arenaceous benthonic forams. Using foram data from modern seas (see Fig. 5-7), this section is interpreted as recording a major marine transgression starting with the Graneros Shale up through the middle of the Greenhorn Formation, and then a subsequent regression on through the Carlile Formation. (After D. Eicher, 1969.)

determine general changes in water depth at one time across a region, or with time in any one place (Fig. 5-8). Of course, the more such depth indicators we can use, the more reliable the interpretation.

Besides seeing systematic variations in microfossils and larger shelly invertebrates, we can also observe variations in the kinds of burrows, tracks, and traces—so-called *trace fossils*—across broad environmental regimes. Even though we often cannot identify the organism responsible for a specific trace fossil, we can usually interpret the behavior that such a fossil represents: dwelling, feeding, crawling, resting, and so on. In addition, by studying the distribution of modern traces in different environments, we can assign a general

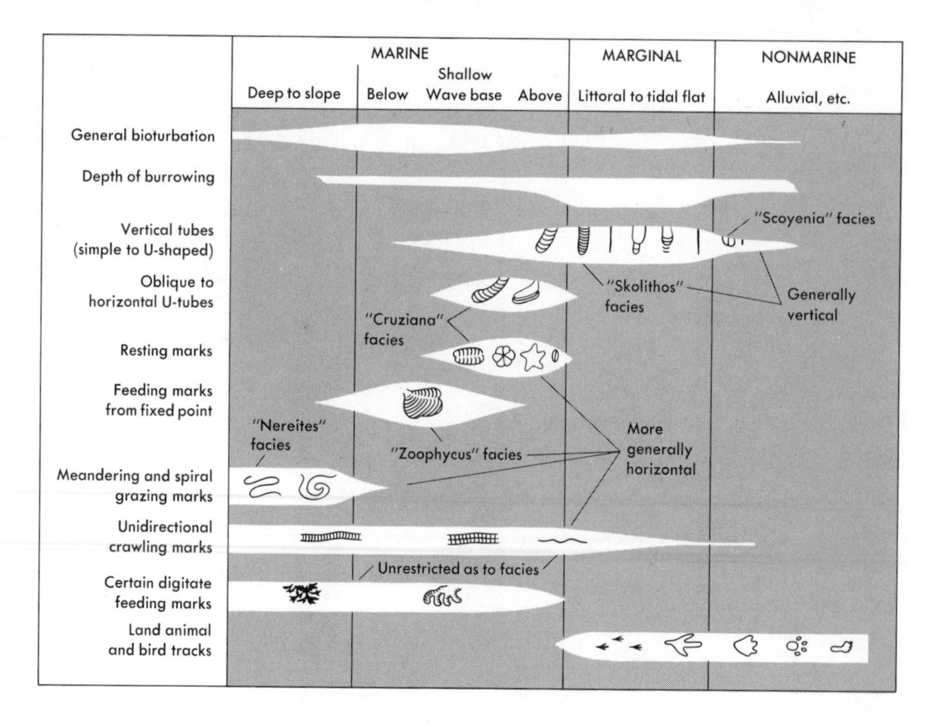

FIG. 5-9 Distribution of various kinds of trace fossils in modern sedimentary environments. Note variations in degree of burrowing—or bioturbation—and depth of burrowing as well as the direction of burrowing (horizontal or vertical). (After Heckel, 1972.)

environmental label to ancient traces. The various kinds of trace fossils seen in modern seas and nonmarine environments are shown in Fig. 5-9.

Burrows within sediments tend to be increasingly abundant in quieter water for two reasons. First, fine-grained sediments accumulate in quiet water (and with them, particulate organic matter) and provide a rich food source for deposit feeders. Second, the lack of repeated reworking of the burrowed sediments by waves and currents preserves the traces. In nearshore tidal flats, burrows are often found to be vertically directed, because the infauna that makes them needs to be well into the sediment to prevent desiccation during low tide and consequent subaerial exposure. Horizontally-directed burrows are more common in offshore waters where there are high populations of infaunal deposit feeders. Here, too, are found surface tracks and trails made by a wide variety of crawling scavengers and browsers. Certain of these feeding traces are more likely to be in one area than another, depending on the ecological requirements of the organism that made the trace. There is now a large literature on the environmental significance of many different kinds of traces, and paleoecologists find them as useful as the remains of hard-part secreting organisms in interpreting ancient environments (as we shall see in the next chapter).

GEOCHEMICAL ENVIRONMENTAL EVIDENCE

We now want to review how geochemical evidence can be used for environmental analysis. The question we will be posing is this: What kind of chemical evidence can be obtained from sedimentary rocks and their included fossils that will tell us something significant about the environment in which they formed?

Although geochemical environmental analysis has received considerable attention recently, there is not as yet any generally reliable body of geochemical data or a set of procedures that gives unequivocal environmental evidence. The reason is that, although a given chemical system may be in equilibrium with a particular sedimentary environment (thereby perhaps uniquely characterizing that environment), postdepositional changes in environment establish new chemical equilibria. Thus, even small changes in temperature, pressure, and water content and mobility in the interstices of a sediment, can result in significant changes in the chemical attributes of a sediment or rock. The new, postdepositional chemistry will mask or replace the original chemistry that might have characterized the depositional environment. Indeed, the chemical attributes of a sediment that might identify a specific environment are usually far more susceptible to postdepositional alteration or obliteration than are fossils or the physical features of a sediment. Of course, chemical—and, by extension, mineralogic—evidence will often record and explain the postdepositional history of a rock. Yet, despite these inherent difficulties, sedimentary geochemistry has considerable promise in identifying and interpreting original environments.

Many calcium carbonate-secreting marine algae and invertebrates form different mineral varieties of $CaCO_3$—namely, aragonite or calcite—depending on local water temperature (Fig. 5-10). Some invertebrates, such as various clams, snails, bryozoans, and serpulid worms, will secrete $CaCO_3$ both as calcite and aragonite, but will vary the ratio with changing water temperatures. Thus, in cold water certain forms will have high calcite-to-aragonite ratios, whereas in warmer waters the same species will have lower calcite-to-aragonite ratios.

Other groups of organisms, such as stony reef-building corals, which secrete only aragonite, will have many warm-water species but only a few

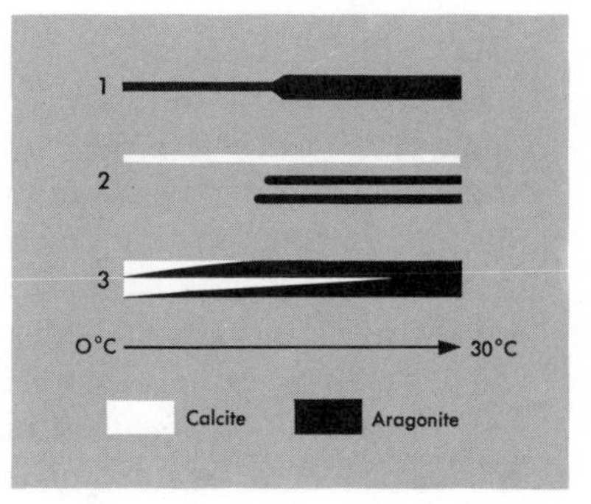

FIG. 5-10 Schematic diagram showing three major kinds of calcite-aragonite variations among marine calcareous algae and invertebrates. Group 1 includes organisms that secrete only aragonite, but that have more species in warmer waters than in colder waters (for example, reef-building corals). Group 2 includes organisms that secrete both calcite and aragonite, although the aragonite species are restricted to warm waters (for instance, certain algae and alcyonarian coelenterates.) Group 3 includes organisms that individually secrete both calcite and aragonite in the same shell but with the ratio charging with temperature (such as many clams, snails, bryozoans, and serpulid worms). (From H. Lowenstam, 1954.)

cold-water varieties. Certain aragonite-secreting red algae and alcyonarians (a group of colonial coelenterates) will occur only in warm water, while closely related, calcite-secreting species will be found in both cold and warm water.

Although the chemical dynamics are not fully understood, it is clear that the secretion of $CaCO_3$ in the form of the mineral aragonite is definitely favored in warmer marine waters. Therefore, an increase in the abundance of aragonitic fossils in an ancient rock would, other things being equal, indicate increasing water temperatures of the water in which those organisms once lived. But "other things" are not usually equal. For example, the mineral aragonite, generally, is not stable, and with time it spontaneously recrystallizes into the more stable crystal structure of calcite. Rocks older than Cretaceous age, therefore, contain few aragonitic fossils. Paleozoic rocks, with one or two exceptions, lack aragonitic fossils altogether. Consequently, this temperature relationship between mineral and water is only potentially useful for Cretaceous and younger fossils.

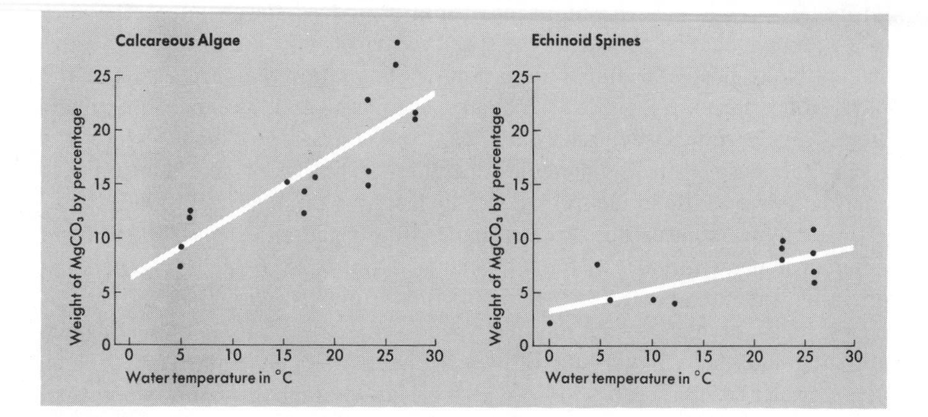

FIG. 5-11 The increase of magnesium in the mineral calcite with increasing water temperature, in calcareous algae and echinoid spines. Magnesium substitutes for calcium in the crystal lattice of the mineral calcite, $CaCO_3$, forming $MgCO_3$. Note that the amount of magnesium substitution is greater for the calcareous algae than it is for echinoids over the same temperature range. (From K. Chave, 1954.)

With increasing water temperatures, many marine algae and invertebrates that secrete calcite will vary the amount of magnesium, which substitutes for some of the calcium in the calcite-mineral structure. Within a given group of organisms, higher magnesium values are found in the warmer-water part of the organism's range than in the colder-water part (Fig. 5-11). But here too this relationship has been hard to establish in ancient fossils, for high-magnesium calcite is very unstable and after burial quickly converts to low-magnesium calcites as the magnesium is leached out by water in the surrounding sediments.

Organically-secreted calcium carbonate may contain other clues, besides mineralogic ones, regarding the temperature of the surrounding water. The element oxygen—which is, of course, a constituent of calcium carbonate ($CaCO_3$) —has several different isotopes. These isotopes, which are chemically identical varieties of oxygen though they have different atomic masses, are designated as oxygen 16, 17, and 18, depending on the total number of protons and neutrons in the nucleus.

Although these isotopes have relative abundances of 99.76, 0.04, and 0.20, (O^{16}, O^{17}, and O^{18}) respectively, these relative abundances vary slightly but perceptibly with temperature. Thus, the O^{18}/O^{16} ratio of ocean water varies indirectly with increases in water temperature. For example, sea water at 10°C will have a relatively higher O^{18}/O^{16} isotopic ratio than sea water at 30°C (Fig. 5-12).

FIG. 5-12 (A) Decrease in the relative abundance of O^{18} with increasing water temperature $R_s = O^{18}/O^{16}$ of sample; $R_o = O^{18}/O^{16}$ of standard. (B) Temperature variations observed in Mesozoic belemnite (an extinct cephalopod) based on oxygen isotopes. Periodic high and low temperatures reflect summer warming and winter cooling of the sea during the life of the belemnite. There is also a suggestion that the surrounding average water temperature declined as the animals aged, indicating that the belemnite migrated to cooler waters or lived at greater water depths. (C) The calcareous skeleton of a belemnite of Jurassic age from North Dakota. (D) Cross-section of belemnite, showing darker and lighter winter (W) and summer (S) growth layers. Annual layers are numbered. (From M. Kay and E. Colbert, 1965.)

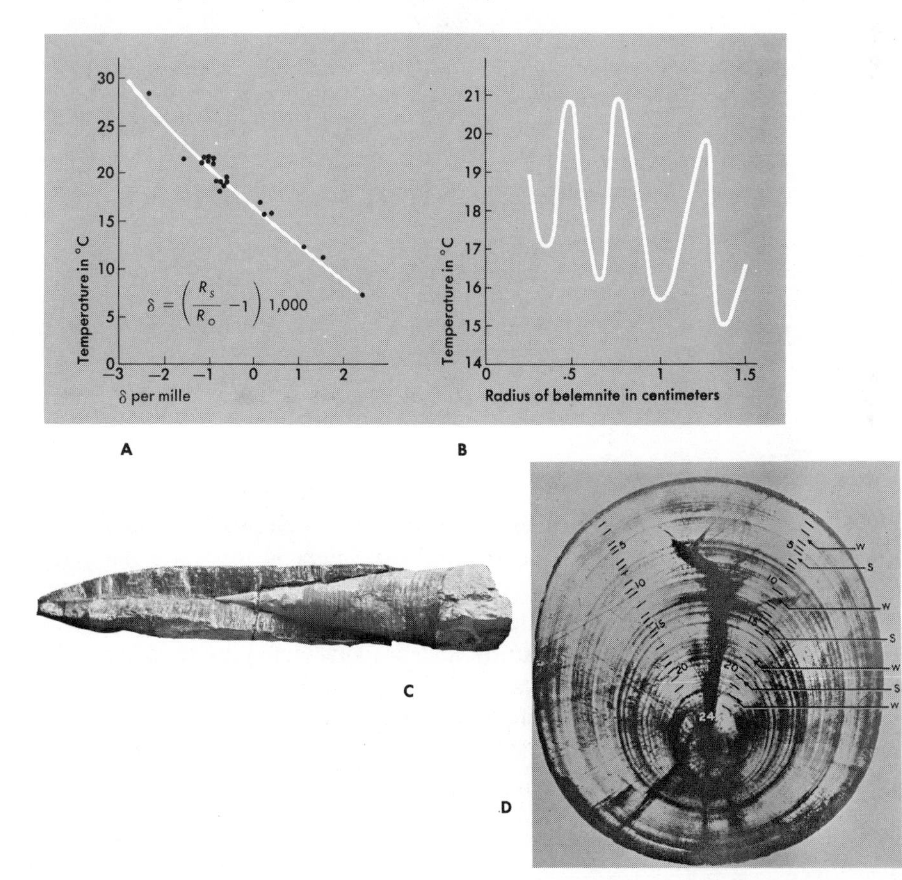

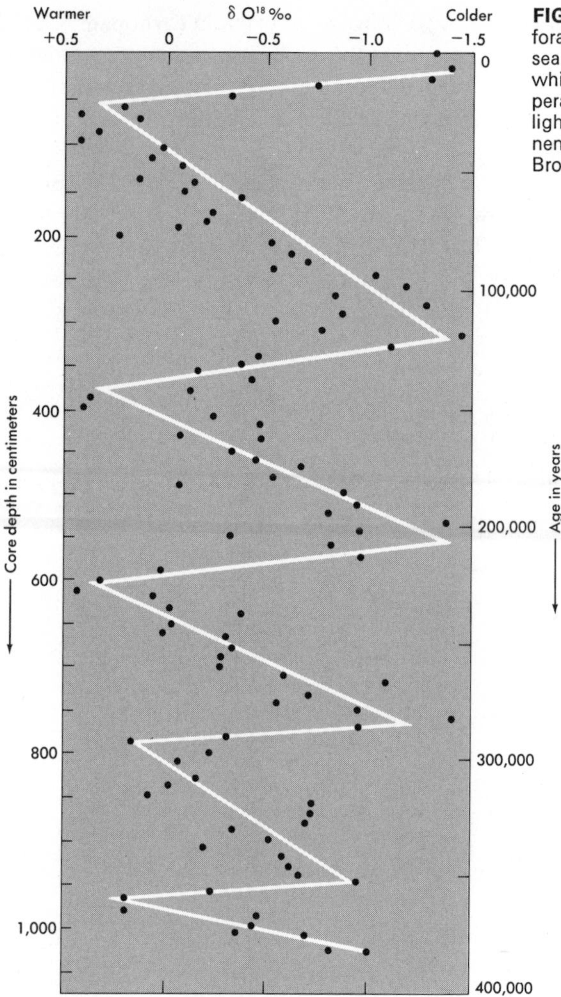

FIG. 5-13 Variations in O^{18} in foram tests going down a deep-sea core from the Caribbean Sea, which reflect changes in sea temperature and amount of isotopically light water contained in continental glaciers. (After W. S. Broecker and J. Von Donk, 1970.)

There is also evidence that during glaciations the oceans had a higher O^{18}/O^{16} ratio, because as water evaporates from the ocean surface the heavier oxygen isotopes tend to be left behind and the lighter isotopes are relatively more abundant in snow and ice that form continental glaciers. Consequently, during the glacial ages ocean water was isotopically heavier with respect to oxygen owing to lowered temperatures and proportionately more isotopically light water tied up in continental glaciers. As shown in Fig. 5-13, oxygen isotope variations in the calcareous tests of planktonic foraminiferans in deep sea cores record glacial and interglacial periods over the last several hundred thousand years.

Other isotopes can provide additional information about ancient environments. For example, the C^{13}/C^{12} ratio found in land plants varies considerably from that of marine plants. Shells, too, secreted in sea water have different ratios of carbon isotopes from those of shells secreted in fresh water (Fig. 5-14). Determination of C^{13}/C^{12} ratios in ancient rocks and fossils, therefore, should suggest the nature of the original environment in which these sediments were laid and fossils formed, if this wasn't already apparent in the first place.

An interesting use of both carbon and oxygen isotopes to trace the environmental changes in Lake Erie near the end of the last glaciation is shown in Fig. 5-15. The ratios come from freshwater ostracods and molluscs. The relatively high C^{13} values for the ostracods, some 13,000 years ago, means that the dissolved carbon in the lake water was close to that of atmospheric carbon dioxide, with little or no aquatic vegetation present, which would have otherwise added isotopically light carbon to the lake water upon decay (refer back to Fig. 5-14 to see that this is so). About 8,000 to 13,000 years ago, the climate ameliorated with an accompanying increase in aquatic vegetation and lighter carbon isotopes. The higher O^{18} values also suggest a warming trend at this time, which continued to about 6,000 years ago. The reversal in the carbon-isotope ratios from 8,000 to 6,000 years ago, during the warming trend, may reflect increasing water depth of the lake and, therefore, relative decrease in the contribution of shallow, aquatic vegetation to the carbon-isotope reservoir.

The reasons for the changes in oxygen-isotope ratios in the last several thousand years are not certain, but may be due to decreased evaporation with lake deepening, changes in drainage patterns of the Great Lakes, and so on. In any event, the approach illustrated in Fig. 5-15 is an interesting one and suggests how stable isotopes may be used for analyzing ancient environments, particularly in conjunction with other nongeochemical evidence.

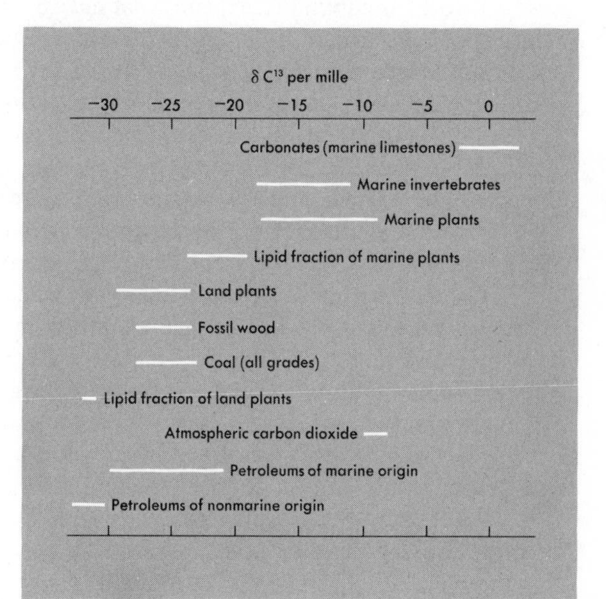

FIG. 5-14 Variations in carbon isotope ratios (C^{13}/C^{12}) in natural materials from different environments. Note that land plants, fossil wood, and coal all have about the same ratio, thereby corroborating other evidence that coal is formed from land plants. The origin of petroleum, on the other hand, apparently derives from only the lipid portion, or fats and waxes, of marine and nonmarine plants. (From R. M. Garrels and F. T. Mackenzie, 1971.)

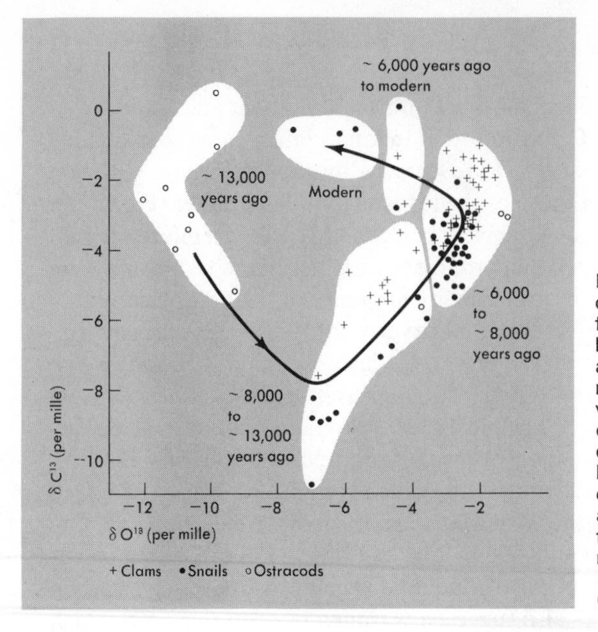

FIG. 5-15 Changes in carbon and oxygen isotopes in calcareous shells from Lake Erie sediments for the last 13,000 years. Relative decrease and increase in C^{13} over that time record an increase in aquatic vegetation with improvement in climate, and then a subsequent decrease with deepening of the lake. Relative decrease and increase of isotopically heavy oxygen reflect a general warming trend and variations in evaporation rates, in part related to lake deepening. (After P. Fritz, T. W. Anderson, and C. F. M. Lewis, 1975.)

SUMMARY

Environmental analysis of sedimentary rocks and their included fossils leading to paleoecological inferences has a basic strategy all its own. Because we cannot observe directly the interactions of fossils with their environments, we must use our knowledge of the ecological relationships of living organisms to guide our inquiry into the past. Using a modern analogue for our ancient examples usually clarifies important organism–environment interactions that we would not otherwise discover if we concentrated only on the fossils and the rocks.

Before attempting to explain what environmental factors limit the distribution and abundance of the fossils themselves, we (or someone else) must first work out the environmental stratigraphy of the sedimentary unit in which the fossils occur. This includes definition, delineation, and interpretation of the facies; recognition of an environmental datum; and prediction of possible ecologic gradients across the strike of the facies. Some gradients, like those of temperature, oxygen, and dissolved nutrients, are difficult to establish directly, but they may be related to other, more easily recognizable gradients, such as water energy, substrate texture, and changes in taxonomic abundances.

Geochemical evidence can also provide evidence about past environments, although postdepositional chemical alteration may severely obscure it. Variations in calcite–aragonite ratios of marine shells as well as variations in magnesium content or ratios of stable isotopes of oxygen and carbon reflect changes in environmental parameters, especially water temperature.

No one line of evidence, however, will usually be enough to reconstruct the ancient environment. Rather, multiple lines of independent evidence must be woven together, logically and consistently, to create a cogent fabric of paleoecological interpretation.

six

environmental synthesis

Our task, then, is to identify the remains that lived together, reconstruct . . . the community structure . . ., and infer its ecological and evolutionary significance. (James W. Valentine, 1973)

Having discussed the interactions of organisms, sediments, and environments, how they are displayed (or not displayed) in the stratigraphic and fossil record, and some of the assumptions and methods of environmental analysis, we now turn to three particular examples to indicate how, in fact, paleoecologists read out ancient environments and what their general significance might be for historical geology. The examples are chosen from widely different parts of the stratigraphic record: marine and terrestrial rocks, carbonates and clastics, macrofossils and microfossils, invertebrates and vertebrates, Paleozoic and Cenozoic times.

The first example comes from Devonian limestones that were deposited at the edge of a transgressing sea in central New York. Based upon our knowledge of modern carbonate environments, the physical characteristics of these limestones indicate deposition in tidal-flat and shallow subtidal environments, with varying degrees of water energy and circulation. The biofacies associated with these environments can then be interpreted in terms of their major ecologic controls. This example from the Devonian Period indicates the usefulness of working out the environmental stratigraphy before interpreting the paleoecology of the fossils. Moreover, the model of clear-water, epeiric sedimentation fits this example very well. After describing and interpreting these early Devonian limestones, we will compare them with middle Ordovician limestones of similar aspect to see how their respective fossil communities resemble each other, and what significance such resemblance might have.

Our second example is from early Permian deltaic rocks of north central Texas. Here we find a water-adapted ecosystem whose vertebrate fauna ranges from ponds and streams to flood plains and upland divides that lasted for several million years, until the aridity increased to a point where the ecosystem died out. Because many members of this ecosystem depended upon freshwater for their existence—fishes and amphibians, especially—other members like the reptiles that fed upon them also remained water-bound, even though they were otherwise fully adapted to terrestrial conditions. When we compare this early Permian ecosystem to a later one in early Triassic time, we see a major shift away from dependence on water-adapted prey to terrestrial herbivores. Unlike the Ordovician–Devonian comparison of nearshore invertebrate marine communities, communities of terrestrial vetebrates do show evolution in their community structure.

Our final example from the Pleistocene sediments of the North Atlantic provides a completely different paleoecological approach from the previous two examples. Here we use the modern ecology of planktonic foraminiferans and other microfossils to read out the marine conditions that existed hundreds of thousands of years earlier. Initially, this can be done qualitatively: individual cold- and warm-water species in deep-sea cores used to interpret cold and warm intervals during Pleistocene time (refer back to Fig. 5-13). However, it also seems possible to make quantitative estimates of past water temperatures and salinities by using known temperature–salinity tolerances of extant Pleistocene species and oxygen-isotope ratios. In this way it becomes feasible to contour the Pleistocene sea surface with respect to temperature and salinity. As such details of past oceans are clarified, we may then draw conclusions about the future oceans and the climatic conditions responsible for them. Thus, as the present is a key to the past, the past in turn may become a key to the future.

EARLY DEVONIAN SEA OF NEW YORK

Throughout the early and middle Paleozoic Era, eastern North America was repeatedly covered by shallow epeiric seas in which a variety of fossiliferous sediments accumulated. Orogenies in late Ordovician and late Devonian times created what we now call the Appalachian Mountains. Uplift and erosion of these ancient mountains produced thick wedges of terrigenous clastics along their flanks, ranging from deltaic redbeds to nearshore sandstones and offshore siltstones and shales. Thus, from Cambrian through Devonian rocks, there is an excellent record of life and environments spanning almost one-quarter billion years of Earth history.

Helderberg Fossils and Environments

Across central New York and down along the Hudson River Valley, early Devonian marine limestones and shales record a regional transgression of a shallow sea, northward and westward, through West Virginia, western Maryland, central Pennsylvania, and into central New York. These rocks form the Helderberg Group, which comprise a sequence of limestone and shale formations, with each formation containing distinctive fossil assemblages and rock types (Fig. 6-1). Because each Helderberg formation could be differentiated from another, particularly on the basis of its fossil content, early geologists interpreted this to mean that each of the formations was deposited uniformly across the whole area. Differences in their respective fossils were viewed as the simple result of evolution over the time during which the rocks were deposited.

More recently, however, careful field stratigraphy demonstrated that these formations are better interpreted as the accumulated record of a series of laterally migrating, shallow-water environments accompanying regional transgression. Hence, the *vertical* Helderberg stratigraphy represents the *lateral* biofacies and lithofacies of this shallow sea of early Devonian time (refer to Fig. 6-1, and recall the discussion of Walther's Law in Chapter 2).

FIG. 6-1 (A) The Helderberg Group of marine limestones and shales in central New York. (B) At one time, each formation was thought to be the same age throughout the region, owing to the distinctive fossils of each formation. (C) Later detailed field work showed that the formations interfingered with one another and recorded a series of sedimentary facies, migrating laterally and building up vertically over time. The distinctive fossils of each facies reflect, therefore, different marine environments rather than significantly different geologic ages. Sr denotes the Rondout Formation of the late Silurian Period; Dm, Dc, Dk, and Dns are the Manlius, Coeymans, Kalkberg, and New Scotland formations of the early Devonian Period; Dor is the Oriskany Sandstone, also early Devonian, that lies unconformably on the Helderberg Group, owing to pre-Oriskany/post-Helderberg erosion. (After Laporte, 1969.)

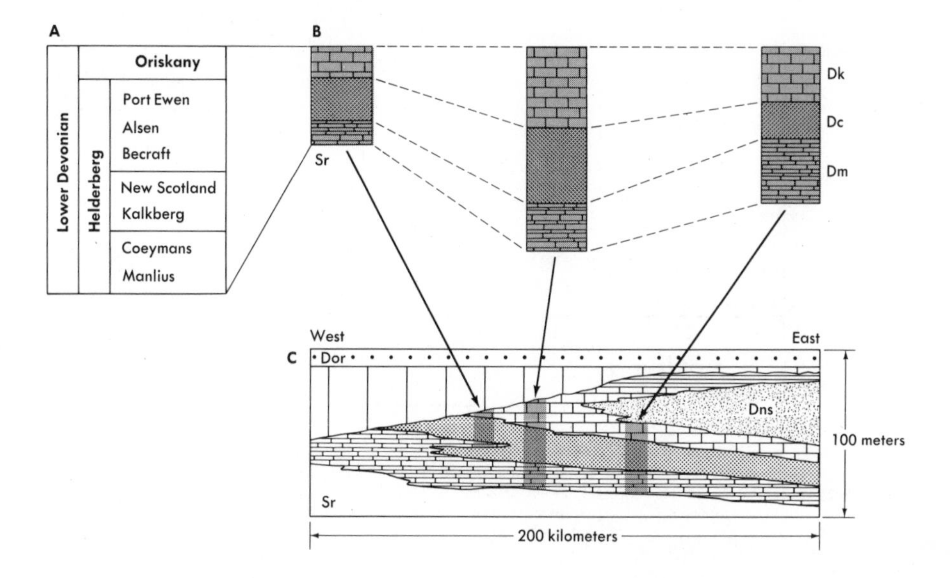

Given this three-dimensional geometry, we can examine the middle portion of the Helderberg Group across the west–east length of outcrop to observe the lateral variations in the lithology, paleontology, and primary structures of these different facies to see if we can reconstruct the original depositional environments. We might think of this as if we were rowing a boat across the ancient Helderberg sea, taking bottom samples of the marine life and sediments on the sea floor, and then attempting to explain what environmental factors were important in controlling the distribution and abundance of the biotas and sediments. Table 6-1 and Figs. 6-2 and 6-3 summarize the results of this lateral facies analysis.

Table 6-1 Lateral Facies Variations in the Helderberg Group

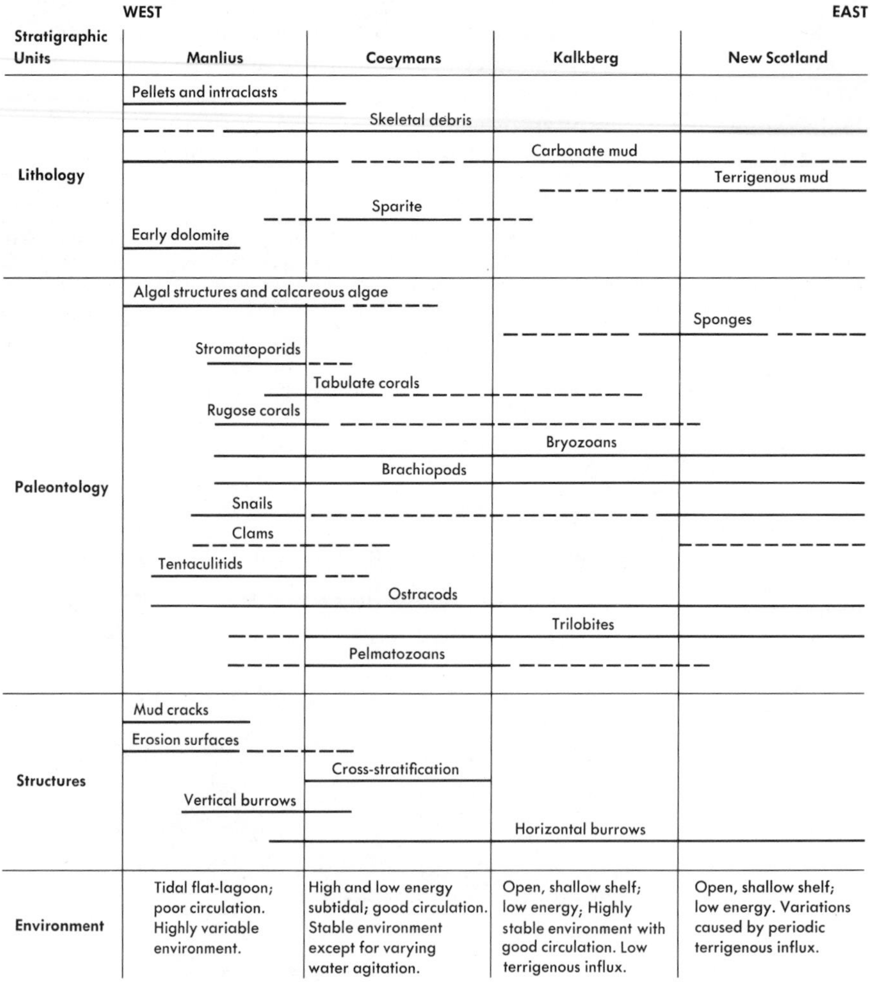

	WEST			EAST
Stratigraphic Units	Manlius	Coeymans	Kalkberg	New Scotland
Lithology	Pellets and intraclasts			
		Skeletal debris		
			Carbonate mud	
				Terrigenous mud
		Sparite		
	Early dolomite			
Paleontology	Algal structures and calcareous algae			
				Sponges
	Stromatoporids			
		Tabulate corals		
	Rugose corals			
			Bryozoans	
		Brachiopods		
	Snails			
	Clams			
	Tentaculitids			
		Ostracods		
			Trilobites	
		Pelmatozoans		
Structures	Mud cracks			
	Erosion surfaces			
		Cross-stratification		
	Vertical burrows			
			Horizontal burrows	
Environment	Tidal flat-lagoon; poor circulation. Highly variable environment.	High and low energy subtidal; good circulation. Stable environment except for varying water agitation.	Open, shallow shelf; low energy; Highly stable environment with good circulation. Low terrigenous influx.	Open, shallow shelf; low energy. Variations caused by periodic terrigenous influx.

(*After Laporte, 1969.*)

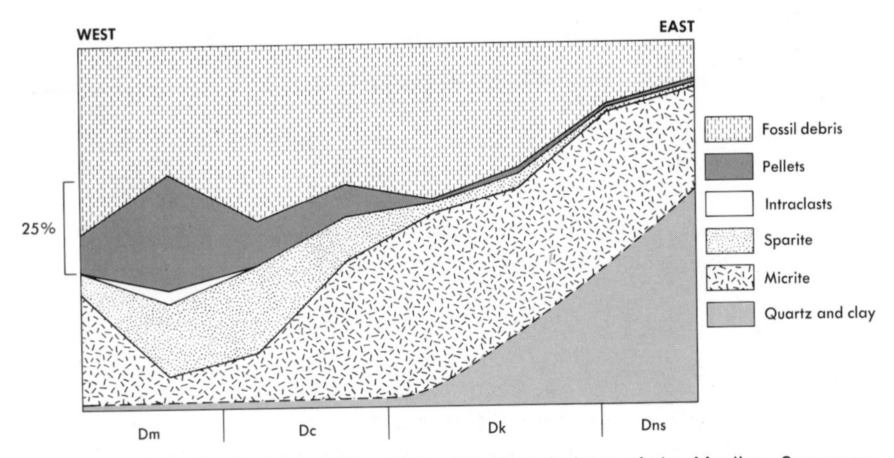

FIG. 6-2 Variation in the composition of the lateral equivalents of the Manlius, Coeymans, Kalkberg, and New Scotland formations, which are limestones and calcareous shales. Pellets include both fecal pellets and small erosional carbonate clasts; intraclasts are larger erosional clasts; sparite is secondarily precipitated calcite within interstices of original grains; micrite is fine-grained, recrystallized carbonate mud; clastic components include quartz silt and clays. (After Laporte, 1969.)

How then are we to interpret these rocks and fossils? First, the presence of mud cracks and local erosion surfaces (scour-and-fill structures) with associated intraclasts indicates intermittent subaerial exposure of the marine sediments that make up large portions of the Manlius Formation. We know that these are marine sediments, owing to the scattered remains of marine organisms, like brachiopods, bryozoans, and corals. This, then, provides us with a valuable environmental datum upon which we can hang our interpretation of the other, related facies, or formations.

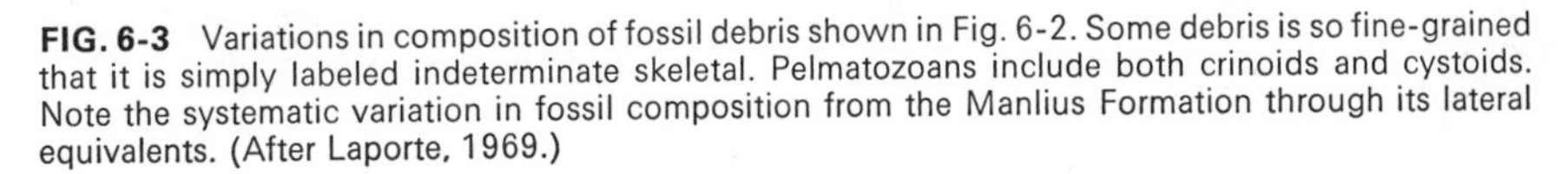
FIG. 6-3 Variations in composition of fossil debris shown in Fig. 6-2. Some debris is so fine-grained that it is simply labeled indeterminate skeletal. Pelmatozoans include both crinoids and cystoids. Note the systematic variation in fossil composition from the Manlius Formation through its lateral equivalents. (After Laporte, 1969.)

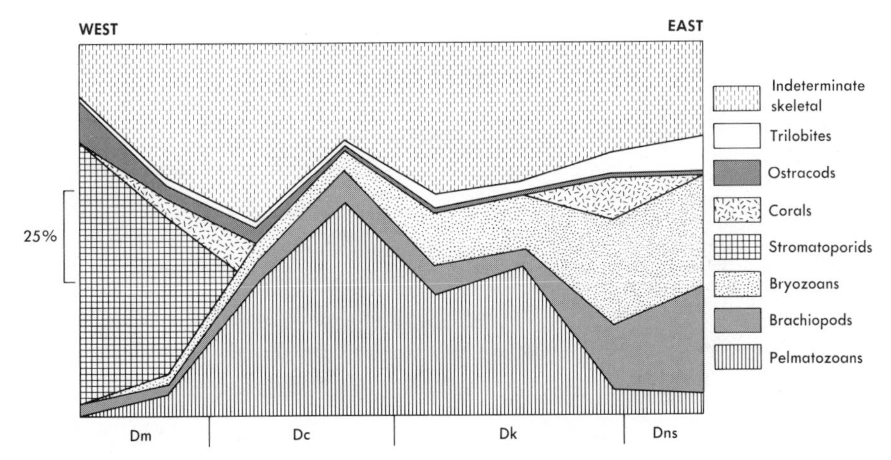

We can recognize an important ecologic gradient paralleling the rapid increase in diversity of marine fossils as we proceed from the Manlius Formation eastward into the laterally equivalent units. Whereas the tidal-flat portion of the Manlius Formation has five to ten species, subtidal Manlius rocks have 25 to 30, the Coeymans Formation 50 to 80, and the Kalkberg and New Scotland formations over 300 species. Surely this increase in diversity must indicate an increasingly more normal marine environment eastward of the Manlius environment, one in which variations in salinity, nutrients, water circulation, and so on were less frequent and less drastic.

Algal stromatolites in some Manlius facies further support a tidal-flat origin for parts of the Manlius Formation. The presence of calcareous algae in other Manlius strata argue for waters shallow enough to permit penetration of sunlight. Yet these waters must not have had strong, through-going currents, because Manlius rocks have a carbonate-mud, or micrite, matrix.

The next seaward facies, however, is mud-free; the calcareous silt and sand grains of the Coeymans Formation—mostly skeletal debris—are cemented by secondarily precipitated calcite within their interstices. The abundant attached echinoderms, or pelmatozoans, including crinoids and cystoids, lived with their feeding parts elevated off the sea floor by a stalk or stem. Tabulate corals and robust rooted brachiopods also thrived here. Whereas all other Helderberg facies are well-burrowed, the Coeymans is much less so; but it does have good cross-stratification generated by strong currents. In short, then, we can interpret this facies as a more open, high-energy, marine environment.

FIG. 6-4 Cross section of the epeiric sea in central New York during early Devonian time, based upon facies analysis of the Helderberg group and the clear-water, epeiric-sea model of Shaw and Irwin. Numbered depositional environments include tidal flat (1), protected subtidal (2a), open subtidal and above wave base (2b), open subtidal and below wave base without terrigenous influx (3), and open subtidal and below wave base with terrigenous influx (4). Compare this figure with Table 6-1. The vertical scale is greatly exaggerated; water depths were several tens of meters at most, whereas the horizontal dimension of the figure is some 100 kilometers or more. (After Laporte, 1969.)

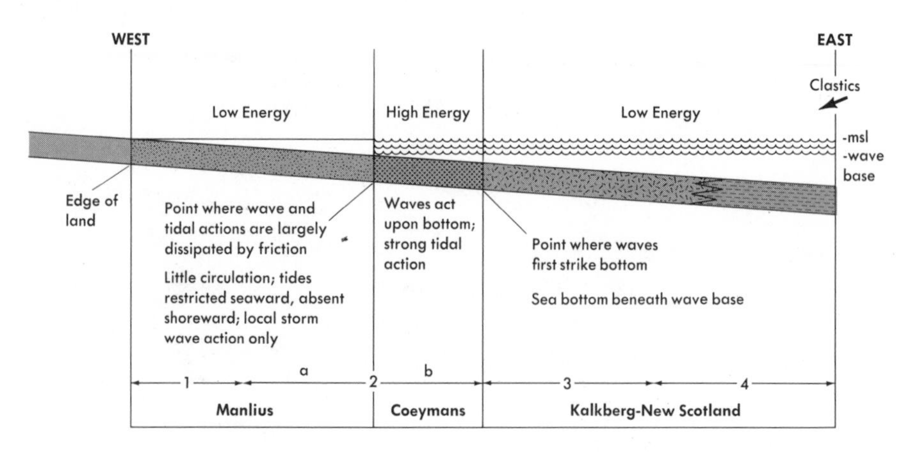

The next two seaward facies of the Kalkberg and New Scotland formations are similar in the abundance and diversity of Devonian marine invertebrates they include, and in their muddy and well-burrowed matrices. However, while the Kalkberg Formation is a quite pure limestone, the New Scotland Formation contains varying amounts of quartz silt and clay, presumably derived from erosion of lands still much farther to the east. The abundant and burrow-mottled mud, lack of primary current structures, and good marine fauna suggest an open, normal marine environment below wave base. As shown in Fig. 6-4, these environmental interpretations match nicely with the clear-water epeiric sedimentation model described in the preceding chapter. If we imagine a transgressing sea, we can visualize how these facies would migrate inland, with each seaward facies being deposited on the more inshore, adjacent facies. Over time, then, the vertical stratigraphic section that we call the Helderberg Group would build up from the laterally equivalent migrating facies.

Comparison of Helderberg and Middle Ordovician Communities

Seventy million years before the deposition of the Helderberg, central New York was covered by a similar shallow epeiric sea in which marine limestones accumulated. Examination of these rocks, the Black River Group of Middle Ordovician time, shows that they have many of the same facies characteristics of the Helderberg Group, especially the Manlius Formation. The resemblances between Manlius and Black River lithofacies apparently record equivalent sedimentologic responses to slow marine transgressions that occurred in middle Ordovician and early Devonian times in New York.

Given these similarities in physical environment, we might expect that the biofacies, too, would be very much alike, owing to similar ecologic controls exerted by the physical environments upon the faunas. And this is, indeed, the case, as shown in Table 6-2. In fact, we can recognize four distinct communities of fossils, each of which is associated with a particular environment along the edge of the transgressing sea: *supratidal*, *high intertidal*, *low intertidal*, and *subtidal*. The limiting factor in this spectrum of environments is the degree of periodic subaerial exposure, ranging from regular and extended exposure in the supratidal to none in the subtidal, where the sea floor and its inhabitants are constantly submerged beneath the sea.

Each community in the Black River and Manlius formations although separated in time by tens of millions of years, has a very similar cast of characters (Fig. 6-5). True, the species and genera are different in each case, but the similarities can be seen at family and ordinal levels. What this suggests, of course, is that the basic adaptations achieved by the various organisms to these environments is recorded in their general morphology, and consequently in their taxonomy, and that these adaptations have remained the same from middle Ordovician through early Devonian time. For example, rather large, smooth-

Table 6-2 Major Taxa and Modes of Life in Black River/Manlius Limestones

Major Taxon	Community Taxa Ordovician Black River Group/ Devonian Manlius Formation	COMMUNITY: Supratidal	COMMUNITY: High Intertidal	COMMUNITY: Low Intertidal	COMMUNITY: Subtidal	RELATION TO SUBSTRATE: Nektobenthonic	RELATION TO SUBSTRATE: Epifaunal Attached	RELATION TO SUBSTRATE: Epifaunal Sessile	RELATION TO SUBSTRATE: Epifaunal Vagile	RELATION TO SUBSTRATE: Infaunal Sessile	RELATION TO SUBSTRATE: Infaunal Vagile	FEEDING TYPE: Suspension Feeder	FEEDING TYPE: Deposit Feeder/ Scavenger	FEEDING TYPE: Browser	FEEDING TYPE: Predator	FEEDING TYPE: Primary Producer
Leperditiid Ostracods	*Leperditia/Herrmannina*	X	X	X					X				X			
Blue-Green Algae	Algal Mats/Algal Mats	X	X	X												X
???	Straight Vertical Burrower/Straight Vertical Burrower		X							X		X				
???	U-shaped Vertical Burrower/U-shaped Vertical Burrower		X							X		X				
Green Algae	Algal Oncolites/Algal Oncolites			X												X

Strophomenid brachiopods	*Strophomena/Mesodouvillina*	X	X			X			X				
Spiriferid brachiopods	*Zygospira/Howellella*	X			X				X				
???	Medium Burrower/Medium Burrower	X						X		X			
???	Large Burrower/Large Burrower	X						X		X			
Ramose Bryozoan	*Stictopora*/Unidentified Trepostome	X			X				X				
Gastropods	*Loxoplocus/" Loxonema"*		X				X				X		
Articulate Codiacean Algae	*Hedstroemia/Garwoodia*		X										X
Filamentous Green Algae	*Girvanella* Oncolites/*Girvanella* Oncolites		X										X
Stromatoporoids	*Stromatocerium/Syringostroma*		X		X				X				
Solitary Rugose Corals	*Lambeophyllum/Spongophylloides*		X		X							X	
Tabulate Corals	*Foerstephyllum/Favosites*		X		X				X				
Nautiloid Cephalopods	*Actinoceras/Anastomoceras*		X	X								X	
Dalmanellid Brachiopods	*Dalmanella/Dalejina*		X		X				X				
Treposome Bryozoan	*Erydotrypa*/Unidentified Trepostome		X		X				X				
???	Small Horizontal Burrower/Small Horizontal Burrower		X					X		X			

(*After Walker and Laporte, 1970*).

FIG. 6-5 Reconstruction of high intertidal and shallow subtidal communities of the Black River and Manlius limestones of New York. Note the close similarity in morphological types in each community, even though different genera are involved. The only major discrepancy is the presence of the small, tubular, bottom-dwelling mollusc, *Tenta-*

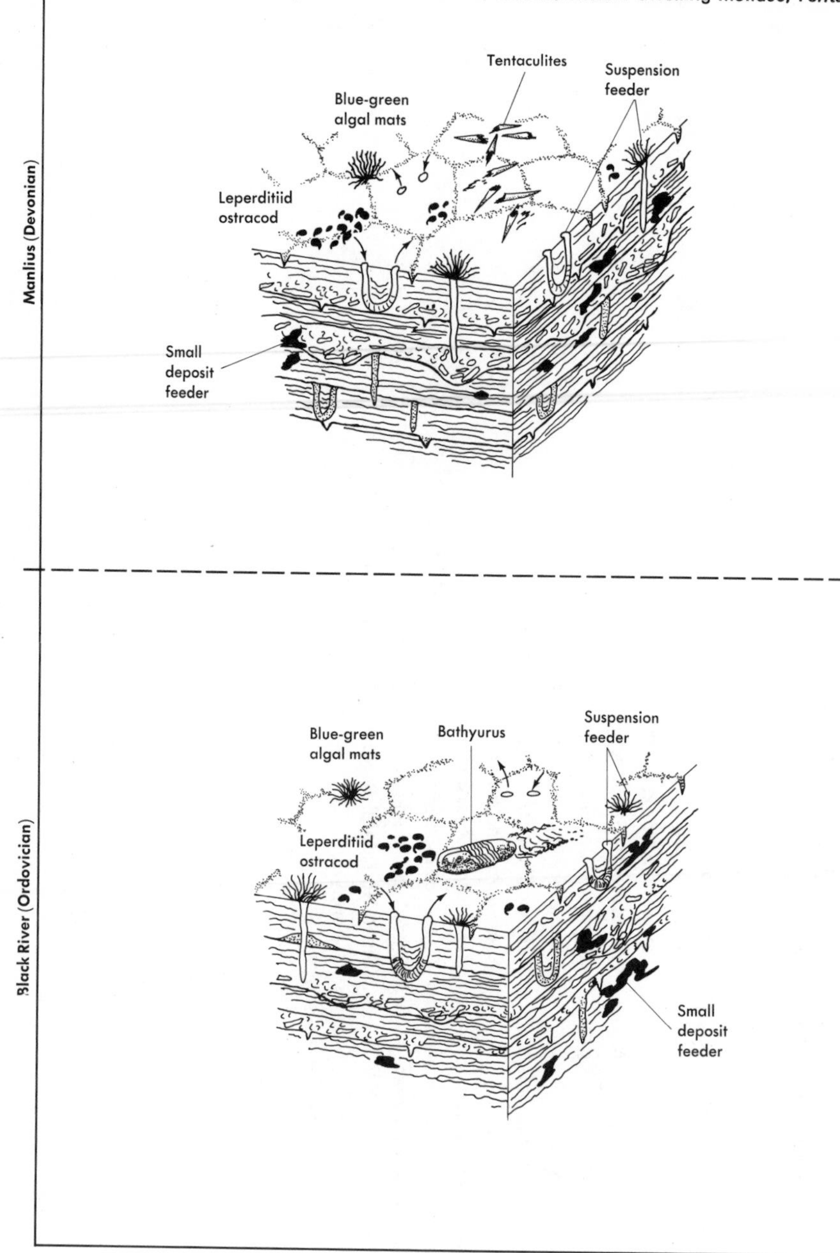

culites, in the Devonian Period, and the bottom feeding trilobite, *Bathyurus*, in the Ordovician Period. One way to explain this difference is to interpret *Tentaculites* as the ecologic replacement of *Bathyurus*. Can you think of other ways to explain this discrepancy? (After Walker and Laporte, 1970.)

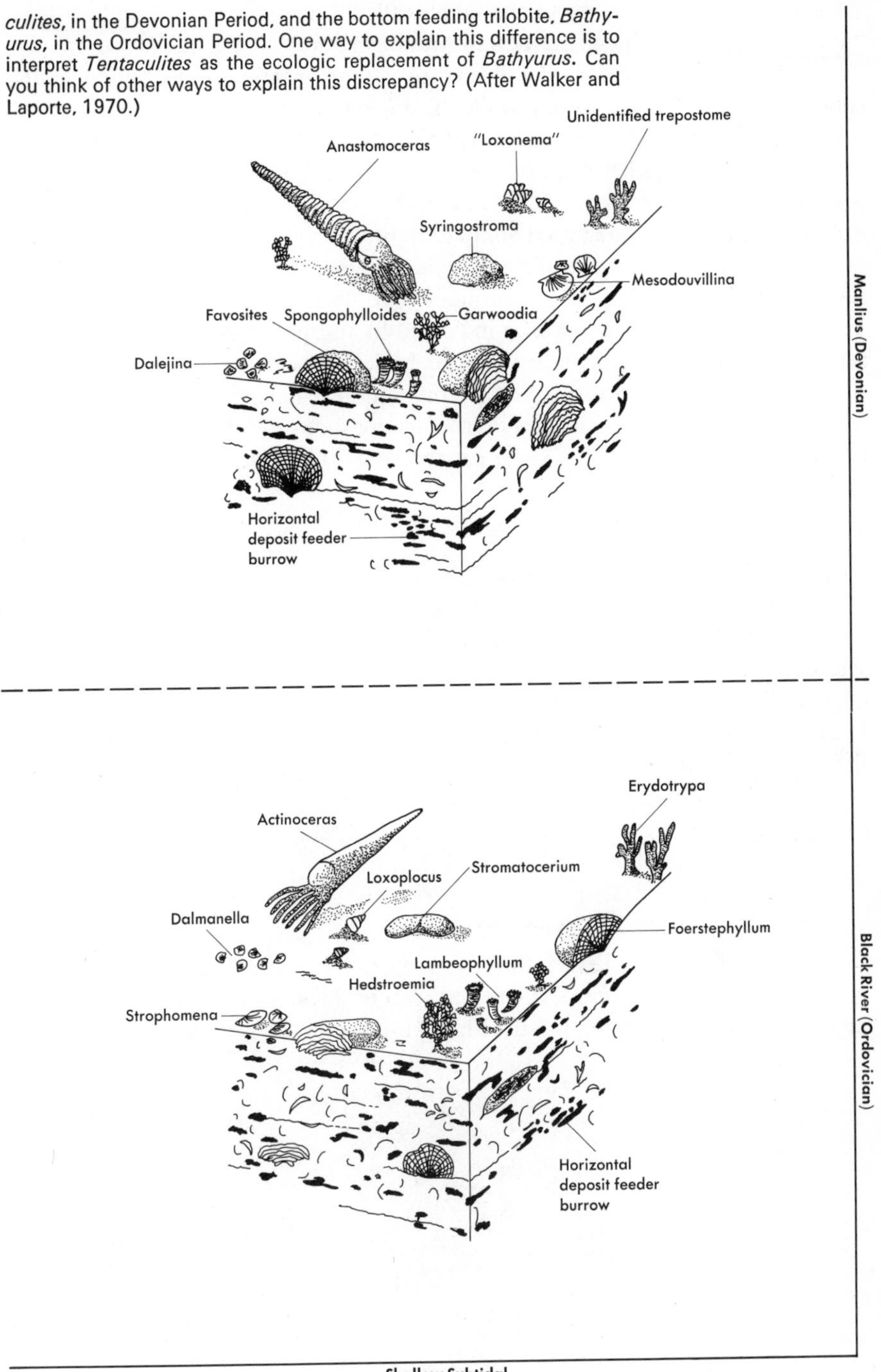

shelled ostracods belonging to the taxon called *Leperditiids* are found in tidal-flat deposits of both Black River and Manlius rocks; the particular genera are, however, different. Or consider the small solitary rugose corals found in the subtidal facies of both units: while belonging to different genera, they are otherwise quite similar in overall morphology and belong to the same taxonomic order. In these and other examples, the taxonomic similarity we are seeing is the result of general morphological similarity, which, in turn, is a preservable sample of the overall integrated adaptation of the organisms to a particular way of life. Of course, at the specific and generic levels, there has been evolutionary turnover and replacement between middle Ordovician and early Devonian time, but the general adaptive types, as indicated by morphology, have remained the same. We may thus conclude that the basic nearshore marine communities found in limestone-depositing seas of the middle Paleozoic Era became established at least soon after the early Ordovician expansion of shelly invertebrates and remained essentially the same for the next 70 million years, into the early Devonian.

EARLY PERMIAN DELTA OF TEXAS

During the late Paleozoic Era, some 250 million years ago, shallow epeiric seas covered much of the southern and western parts of the United States. The edge of the sea was marked by a series of low-lying deltas built from clastic sediments derived from the continental interior. Nonmarine biotas flourished in the various deltaic habitats and included ferns, seed ferns, conifers, and horsetails; freshwater molluscs, worms, and arthropods; freshwater sharks, crossopterygian (lobe-fin) and paleoniscoid (ray-fin) fishes; and various amphibian and reptilian vertebrates.

Stratigraphy and Environments

In north-central Texas, the Clear Fork Group of early Permian age provides an excellent record, some 750 meters thick, of these deltaic environments with their associated fossils. Figure 6-6 shows the general stratigraphy of

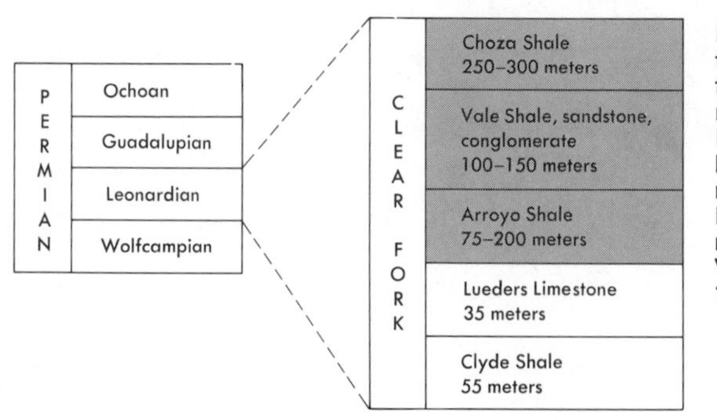

FIG. 6-6 Stratigraphic position of the Arroyo, Vale, and Choza formations of early Permian age in north-central Texas. The Clear Fork Group as a whole records a local marine regression, with the marine Clyde Shale and Lueders Limestone overlain by the non-marine, deltaic shales of the Arroyo, Vale, and Choza. (After Olson, 1971.)

the area and Table 6-3 indicates the depositional environments and inferred topography and rainfall. Within the three nonmarine formations (Arroyo, Vale, and Choza) that make up most of the Clear Fork Group, four major habitats are recognized: *upland*, including flood plains and divides between streams; *streams*, where water movement is relatively great; *ponds and lakes*, where water movement is much less; and *pond and lake margins*, transitional between fully aquatic and nonaquatic habitats.

Each of these four major habitats is identified by physical, inorganic attributes of the rocks, including channeling, channel lag conglomerates, mud cracks, cross-stratification, sediment texture, and three-dimensional geometry of rock types (refer ahead to Table 6-3). Several general biological attributes of the rocks corroborate the interpretation of these habitats, and these include the occasional presence of well-preserved land plants, lungfish burrows and coiled skeletons of aestivating amphibians, and rare freshwater invertebrate fossils.

Within these major deltaic habitats there is a more or less distinct association of fossil vertebrates: (1) purely aquatic forms that include a variety of fish—sharks, ray-finned and lobe-finned bony fish—and certain amphibians; (2) semiaquatic types that lived and fed mostly in the water, but were also able to move about on land or to aestivate by burrowing in temporarily dried-up

FIG. 6-7 Exposure of a Permian pond as seen today in the Arroyo Shale of north central Texas. See Fig. 6-8 for a map of the sediment types and fossils found here. (E. C. Olson UCLA.)

Table 6-3 Environmental Conditions During Deposition of Arroyo, Vale, and Choza Shales*

Formation	Sedimentation	Topography	Rainfall
	Extensive, even clay, sand and evaporite beds. Few channels and ponds. No known life record.	Same	Low rainfall
	↑	↑	↑
Choza 250 to 300 meters	Predominantly flood plains. Channels small, few, widely spaced. Evaporite basins markedly increased. Some playas, freshwater ponds.	Local relief low; land near sea level.	Continued decrease in total rainfall.
	↑	↑	↑
	Few channels, not clay-pebble type. Initiation of evaporite beds. Scattered freshwater ponds.	Local relief moderately low. Basins with fresh-water and evaporite deposits.	Decrease in total; not torrential; seasonality persistent.
Vale 100 to 150 meters	Channel and lag deposits all of clay-pebble conglomerates, sediment locally derived; ponds scattered, predominantly persistent; ponds show seasonality in nature of deposition.	Local relief increased.	"Monsoonal" type fully developed. Torrential seasonal rains.
	↑	↑	↑
	Same, but beginning of clay-pebble conglomerates, derived from local sediments.	Local relief same; source area high.	Initiation of "monsoonal" type.
	↑		

	Marked channel development. Coarse, dipping, marginal flood plain deposits. Few but extensive ponds. Coarse materials of streams derived from source area.	↑ Moderate local relief.	↑
Vale 100 to 150 meters	↑	Marked increase in relief in source area of sediments.	Increase in seasonality in source area of streams.
	Initiation of large stream channels with conglomeratic deposits. Ponds widely spaced.		
	Even red shale, with broad linear belts of silt and fine sand. Deposited by very slowly flowing water. No definite channels and few ponds.	Local relief very low. Area near sea level; possibly tide-water estuaries.	Same.
	↑		↑
Arroyo 75 to 200 meters	Slight increase in ponding, especially temporary ponds.	Same. ↑	Slight increase in seasonality.
	↑		
	Flood-plain deposits predominant, a few scattered ponds. Divides low, streams small.	Local relief low. Source area of streams low.	Moderate, evenly distributed throughout year.

•Read from bottom to top, from older to younger strata.
(*After Olson, 1971.*)

ponds, including lungfishes, labyrinthodont amphibians, and a burrowing, aestivating amphibian (*Lysorophus*); (3) terrestrial animals that lived mainly on dry land but fed mostly in water, including the carnivorous and herbivorous mammal-like reptiles, the fin-back pelycosaurs, and *Diadectes*, an amphibian that probably fed on freshwater molluscs; and (4) terrestrial forms that not only lived on land but also fed there as well, including primitive reptiles (*Captorhinus*), small pelycosaurs, and possibly some amphibians. These habitats and their fossil associations are shown in Fig. 6-7 (page 131) and Fig. 6-8.

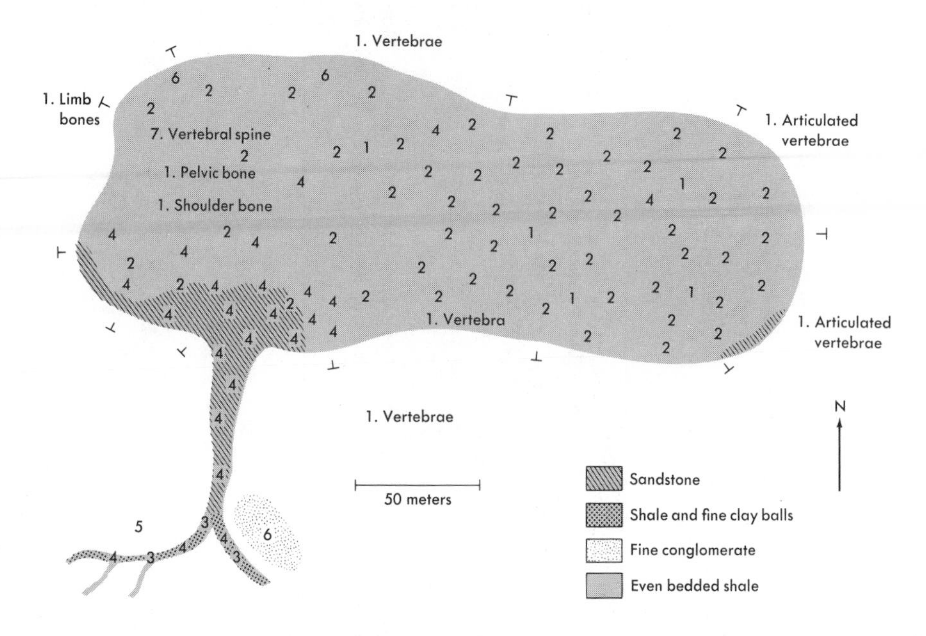

FIG. 6-8 (Above) Map of the area shown in Fig. 6-7, showing the sediment types and fossils in a Permian pond, its shore, and a small stream. Note the dip symbols which indicate that the deposits marginal to the pond dip toward the former water body. Numbers mark the position of the following vertebrates: (1) *Dimetrodon*, a carnivorous pelycosaur; (2) *Diplocaulus magnicornis*, a horned amphibian that lived in the low-energy pond; (3) *D. brevirostris*, another species of horned amphibian that preferred the higher-energy stream channel; (4) Teeth of the shark, *Xenacanthus*; (5) *Diadectes*, a molluscivore amphibian; (6) *Captorhinus*, a terrestrial reptile that fed on insects and worms; and (7) *Eryops*, a semiaquatic carnivorous amphibian. The remains of *Dimetrodon* (1) are partially articulated in the pond margin, but completely disarticulated and fragmentary in the pond itself. Compare with Fig. 6-7. (After Olson, 1958.)

FIG. 6-9 (Facing page) Reconstruction of an early Permian ecosystem based upon fossils and sediments found in the Arroyo Shale, north-central Texas. Feeding relationships and energy flow are shown by arrows; various taxa are placed in their preferred habitats. Relative abundance of organic detritus in each habitat is indicated by the histogram at left. Fishes include *Xenacanthus*, a predatory shark; *Gnathorhiza*, an aestivating lungfish; crossopterygians (lobe-fins) and paleoniscoids (ray-fins). Amphibians include *Lysorophus*, a burrowing aestivator; *Diplocaulus*, *Trimerorhachis*, *Euryodus*, *Trematops*, *Broiliellus*, *Seymouria*, and *Diadectes*. Reptiles include *Edaphosaurus* and *Dimetrodon*, fin-back mammal-like pelycosaurs, and *Labidosaurus* and *Captorhinus*, primitive ancestral reptiles. (After Olson, 1971.)

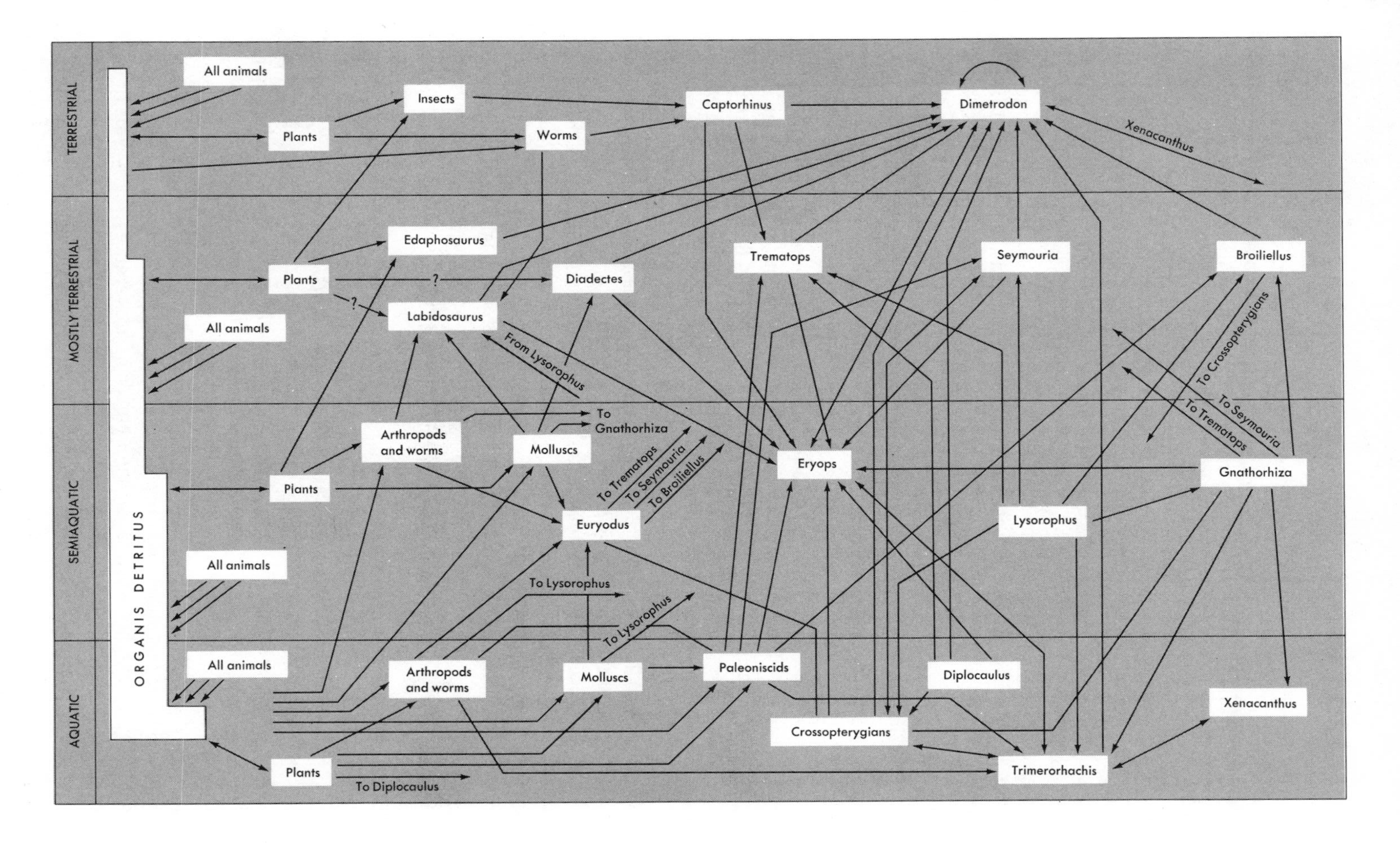
TERRESTRIAL
MOSTLY TERRESTRIAL
SEMIAQUATIC
AQUATIC
ORGANIS DETRITUS
All animals
Insects
Plants
Worms
Captorhinus
Dimetrodon
Xenacanthus
Edaphosaurus
Plants
?
Diadectes
?
Labidosaurus
All animals
From Lysorophus
Trematops
Seymouria
Broiliellus
To Crossopterygians
To Seymouria
To Trematops
Arthropods and worms
To Gnathorhiza
Molluscs
Plants
To Trematops
To Seymouria
To Broiliellus
Eryops
Gnathorhiza
Euryodus
Lysorophus
All animals
To Lysorophus
To Lysorophus
All animals
Arthropods and worms
Molluscs
Paleoniscids
Diplocaulus
Xenacanthus
Crossopterygians
Plants
To Diplocaulus
Trimerorhachis

The fossils and inferred depositional environments of the Clear Fork Group can be reconstructed into a unified ecosystem as shown in Fig. 6-9, in which feeding relationships and habitat preferences of the early Permian vertebrates are schematically indicated. The ecosystem is water-dependent in that much of the organic detritus and nutrients as well as most primary producers and consumers are found in aquatic and semiaquatic parts of the ecosystem, namely ponds, streams, and their margins. Because of this connection to water, even the reptiles, *Captorhinus* and *Dimetrodon*, which could have lived totally in an upland, terrestrial environment, are found within this ecosystem. An approximate modern equivalent of this type of water-based ecosystem is that of the Florida Everglades, with its alligators, snakes, lizards, fish, assorted invertebrates, and varied plants.

Temporal Changes in an Ecosystem

The picture we have just described is typical of the lower Clear Fork formation, the Arroyo Shale. What happens to the fossil assemblages as one moves up through the rest of the Clear Fork Group? Figure 6-10 lists the major vertebrate taxa found in the Arroyo, Vale, and Choza shales of the Clear Fork Group and their preferred habitats. Over the several million years that these formations span, the vertebrate fauna exhibits several significant patterns which we can describe as follows.

First, despite temporal changes in sedimentation, topography, and rainfall (refer back to Table 6-3), there is a persistence of species. Some species, like the fin-back reptile *Dimetrodon gigashomogenes*, and the primitive reptile *Captorhinus aguti*, occur fairly abundantly throughout the upland habitats of the Clear Fork Group. Other species, like the amphibian *Trimerorhachis insignis*, also persist throughout most of the Clear Fork Group, but shift from ponds to streams to ponds again. This shift in habitat preference might be attributed to variations in available food for this omnivore from one habitat to another. Another persistent species is the aquatic, burrowing amphibian *Lysorophus tricarinatus*, that is abundant throughout the Clear Fork Formation in pond deposits where it is found, in most cases, in concentrations of coiled specimens representing mass mortalities of aestivating individuals.

Another interesting modification in the Clear Fork faunas is the apparent evolution of taxa in the younger Vale Formation from *Captorhinus aguti* of the older Arroyo Formation into related herbivores in pond and pond-margin habitats. A different sort of modification concerns the competition presumably exerted on the amphibian pond-dwellers *Diplocaulus* and *Trimerorhachis* shortly after the beginning of Vale deposition by the predatory shark *Xenacanthus*. These two amphibians occurred in ponds during Arroyo time, when the shark was absent, but when *Xenacanthus* made its appearance in early Vale time, the amphibians moved into the higher-energy stream habitat where, apparently, they could survive competition with the shark. The bottom-feeding lungfish

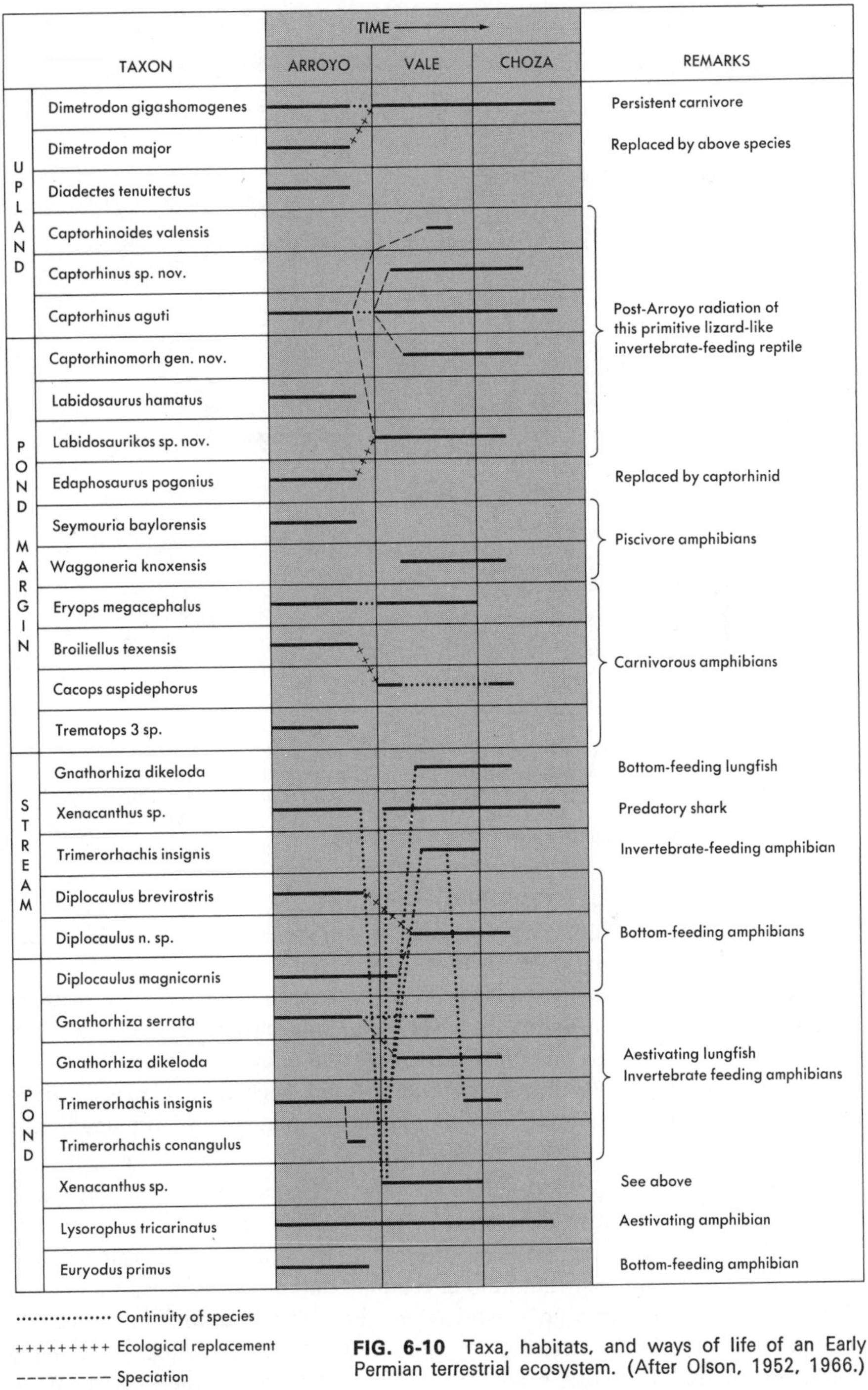

		TIME →			
	TAXON	ARROYO	VALE	CHOZA	REMARKS
UPLAND	Dimetrodon gigashomogenes				Persistent carnivore
UPLAND	Dimetrodon major				Replaced by above species
UPLAND	Diadectes tenuitectus				
UPLAND	Captorhinoides valensis				
UPLAND	Captorhinus sp. nov.				
UPLAND	Captorhinus aguti				Post-Arroyo radiation of this primitive lizard-like invertebrate-feeding reptile
POND MARGIN	Captorhinomorh gen. nov.				
POND MARGIN	Labidosaurus hamatus				
POND MARGIN	Labidosaurikos sp. nov.				
POND MARGIN	Edaphosaurus pogonius				Replaced by captorhinid
POND MARGIN	Seymouria baylorensis				Piscivore amphibians
POND MARGIN	Waggoneria knoxensis				
POND MARGIN	Eryops megacephalus				
POND MARGIN	Broiliellus texensis				Carnivorous amphibians
POND MARGIN	Cacops aspidephorus				
POND MARGIN	Trematops 3 sp.				
STREAM	Gnathorhiza dikeloda				Bottom-feeding lungfish
STREAM	Xenacanthus sp.				Predatory shark
STREAM	Trimerorhachis insignis				Invertebrate-feeding amphibian
STREAM	Diplocaulus brevirostris				
STREAM	Diplocaulus n. sp.				Bottom-feeding amphibians
POND	Diplocaulus magnicornis				
POND	Gnathorhiza serrata				
POND	Gnathorhiza dikeloda				Aestivating lungfish Invertebrate feeding amphibians
POND	Trimerorhachis insignis				
POND	Trimerorhachis conangulus				
POND	Xenacanthus sp.				See above
POND	Lysorophus tricarinatus				Aestivating amphibian
POND	Euryodus primus				Bottom-feeding amphibian

·················· Continuity of species

+++++++++ Ecological replacement

– – – – – – – – – Speciation

FIG. 6-10 Taxa, habitats, and ways of life of an Early Permian terrestrial ecosystem. (After Olson, 1952, 1966.)

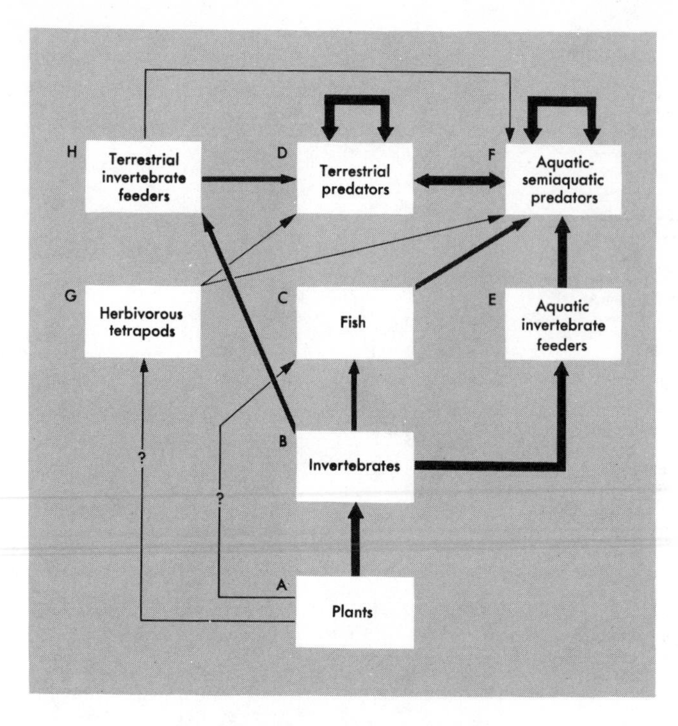

FIG. 6-11 The food chain in Early Permian times. The direction and strength of arrows indicate the relative food flow from one energy level to the next. (A) Aquatic and terrestrial plants. (B) Clams, snails, worms, insects, and other arthropods. (C) Paleoniscoid fish. (D) Carnivorous reptiles. (E) Amphibians. (F) Large amphibians, sharks, and lobe-finned fish. (G) Herbivorous reptiles. (H) Small carnivorous and omnivorous reptiles. (After E. C. Olson, 1961.)

Gnathorhiza, however, not only was able to compete with *Xenacanthus*, but expanded its habitat range along with it.

Finally, one can clearly see from Fig. 6-10 that while the overall ecosystem remained fairly stable throughout much of the nonmarine part of the Clear Fork Group, during early and middle Choza time the ecosystem was eliminated, probably owing to greatly increasing aridity (refer back to Table 6-3). Given that this ecosystem was water-adapted, such an increase in arid conditions would be fatal to it.

Before concluding this section we might compare this Late Paleozoic terrestrial ecosystem with subsequent Early Mesozoic terrestrial communities to see what changes, if any, have occurred (Figs. 6-11 and 6-12). Among other things, it has been noticed that among reptilian and amphibian genera in Late Paleozoic rocks the ratios of carnivores to noncarnivores varied from 4:1 to 7:1. Now this observation is virtually the opposite of what we see today for many terrestrial mammals, where predators are outnumbered by the nonpredators.

If we look more closely, however, at the structure of the Early Permian terrestrial communities—such as the one we have described from the Clear Fork Group of Texas—we find that, although there were, indeed, few genera of terrestrial herbivores, there were a number of aquatic and terrestrial invertebrate feeders. Thus, Late Paleozoic terrestrial carnivores had varied food sources that included not only a few terrestrial herbivores, but also a larger proportion of invertebrate feeders that were water-adapted (fish and especially amphibians), as shown in Fig. 6-11. Gradually, however, a more diverse group of herbivores evolved so that by the Early Mesozoic, some 25 million years later, there was a larger number of terrestrial vertebrate herbivores forming a major part of the diet of the carnivores (Fig. 6-12). This significant alteration in the structure of terrestrial vertebrate communities changes the composition of vertebrate communities from Late Paleozoic to Early Mesozoic time. Moreover, the decreased dependence of these vertebrates on water-adapted forms (aquatic plants → aquatic invertebrates → aquatic and terrestrial invertebrate-feeders) permitted the adaptive radiation of these animals, chiefly reptiles, into a variety of upland, terrestrial habitats. It has been further suggested that the origin of mammals from some of these reptilian groups was initiated by this major change in community structure.

FIG. 6-12 Food chain in the Early Triassic Period. Symbols are the same as in Fig. 6-11. Note the shift in food flow from the aquatic and terrestrial invertebrate feeders to herbivorous reptiles. Primitive insectivorous mammals probably evolved from the terrestrial, invertebrate-feeding reptiles. (After E. C. Olson, 1961.)

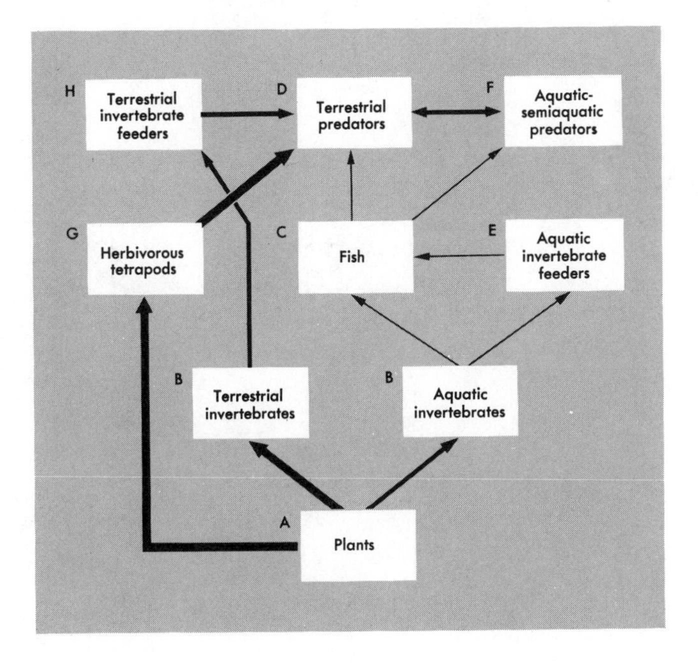

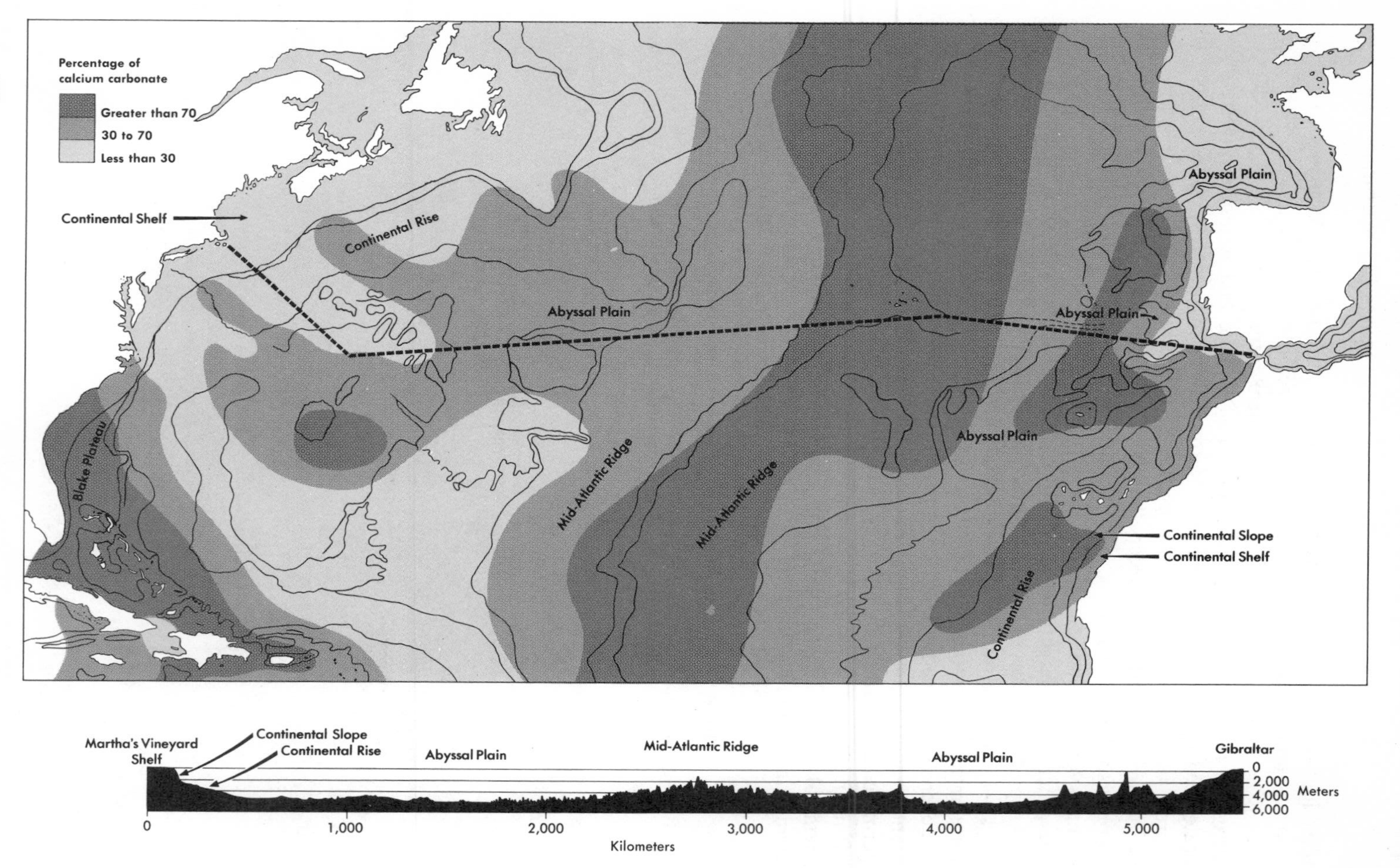
Percentage of calcium carbonate
Greater than 70
30 to 70
Less than 30
Continental Shelf
Continental Rise
Abyssal Plain
Abyssal Plain
Abyssal Plain
Abyssal Plain
Blake Plateau
Mid-Atlantic Ridge
Mid-Atlantic Ridge
Continental Slope
Continental Shelf
Continental Rise
Martha's Vineyard Shelf
Continental Slope
Continental Rise
Abyssal Plain
Mid-Atlantic Ridge
Abyssal Plain
Gibraltar
0
2,000
4,000
6,000
Meters
0
1,000
2,000
3,000
4,000
5,000
Kilometers

LATE PLEISTOCENE EPOCH OF THE NORTH ATLANTIC

The North Atlantic Ocean basin is bounded by the northern portion of South America, the Caribbean islands, eastern North America, the islands of Greenland and Iceland, the British Isles, western Europe, and the northwestern part of Africa. Although in several broad regions the North Atlantic reaches depths greater than 5,000 meters—almost twice that in the Puerto Rican trench—the bottom relief of the basin is quite variable. In fact, the variability of the topographic relief of the North Atlantic basin (and other ocean basins as well) is at least as great as that seen on the continental land masses. But despite this variability, however, certain physiographic provinces are evident (Figs. 6-13 and 6-14).

The basin has formed over the last 200 million years, starting with the breakup of Pangaea, a large supercontinent composed of essentially the present-day dispersed continents, in late Triassic time. According to the theory of plate tectonics, rifting and continuous divergence of the western and eastern portions of the northern part of Pangaea—so-called Laurasia—led to the opening apart of the North Atlantic Ocean; formation of the broad mid-oceanic volcanic ridges; lateral migration of the sea floor from opposite sides of the diverging plate boundary; and sedimentation of a relatively thin layer of pelagic deposits on the underlying submarine volcanics that accumulated along the ridge axis.

Much of the coarser, land-derived detritus that is brought to the ocean accumulates on the continental shelves, which are rather wide areas of little relief that extend seaward from land down to about 200 meters. In addition, turbidity currents will occasionally transport relatively coarse sediments from the shelf edge, or down the submarine canyons that incise the shelf, and deposit them out on the abyssal plains.

Besides these inorganically derived sediments, there are also significant amounts of organic debris accumulating in the deeper parts of the ocean away from the continental shelves. This organic debris comes from a myriad of minute, calcareous tests secreted by floating protozoans, planktonic foraminiferans, as well as planktonic algae, or coccolithophorids. Other, less important skeletal sediments include the tests of planktonic snails (pteropods), various calcareous hard parts of shelly invertebrates, fish bones, and so on.

FIG. 6-13 (Facing page, top) North Atlantic Ocean basin showing major physiographic divisions and general distribution of calcium carbonate sediments. Calcareous-rich sediments are found near ridges and areas of high biological productivity. Calcareous-poor sediments occur in deeper-water areas where calcium carbonate is more easily dissolved. High concentrations around Bermuda and the Bahamas are due to abundant shallow-water shelly invertebrates and calcareous algae.

FIG. 6-14 (Facing page, bottom) Topographic profile across the North Atlantic Ocean along the dotted line shown in Fig. 6-13. Although vertical scale is exaggerated so that slopes appear much greater than they actually are, note the variation in bottom relief.

Qualitative Environmental Changes

With this brief introduction to the present-day North Atlantic basin, let us now consider the paleoecology of the area during the Late Pleistocene Epoch, an interval of time spanning the last 700,000 years or so. Extensive exploration of the oceans began after the turn of this century. Many data were collected on the character and circulation patterns of sea water, the topography and composition of the sea floor, the nature and thickness of deep-sea sediments, and the abundance and distribution of various marine organisms. Particularly important in this oceanographic research was the collection of sediment cores from a large number of stations throughout the North Atlantic. These cores of soft sediment, most of which are 10 meters or more in length, provide a stratigraphic record of deep-sea geologic history. Although most of the cores record only the Holocene and Pleistocene interval, about 10 percent sample pre-Pleistocene sediments, some even going as far back as the Cretaceous Period. These older sediments at or near the sea floor's surface are due to the nondeposition or erosion of younger sediments.

Because of the abundance of planktonic Foraminifera in these cores, and because these single-celled organisms are sensitive to variations in water temperature, it was quickly realized that differences in the foraminiferal species in the cores might provide clues about the ancient temperature of the North Atlantic, particularly during the Pleistocene Epoch, when significant warming and cooling of the Earth's surface occurred.

The scientists studying these cores reasoned that first it would be necessary to determine which planktonic foraminiferans were accumulating in the ocean sediments today, a "warm" interval. Obviously, benthonic foraminiferans would not be helpful, for their distribution, if it is related to temperature, would be controlled by the temperatures prevailing at several thousand meters rather than by surface temperature. And yet surface temperature, of course, would be far more sensitive to major climatic changes than would the bottom waters of the North Atlantic.

By defining the planktonic foraminiferal composition of the uppermost layer of the cores, it would be possible to establish a reference point, or environmental datum, with which other, older assemblages might be compared. For example, one foraminiferan species, *Globorotalia menardii*, is a useful indicator for determining surface temperature because this species, which fluctuates in abundance in the calcareous layers of the cores, is strongly influenced by temperature. In the uppermost layers of the North Atlantic cores *G. menardii* is abundant; going down the core, however, this species disappears for a time, and then reappears once again. This variation in *G. menardii* was attributed to the end of the last glacial age and the beginning of the recent episode of marine sedimentation and climatic amelioration. To judge from the *G. menardii* foraminiferal populations, therefore, a warm, preglacial interval was followed by a cold period (glacial interval) without *G. menardii*, which was followed by another warm period (postglacial) with *G. menardii* again (Fig. 6-15).

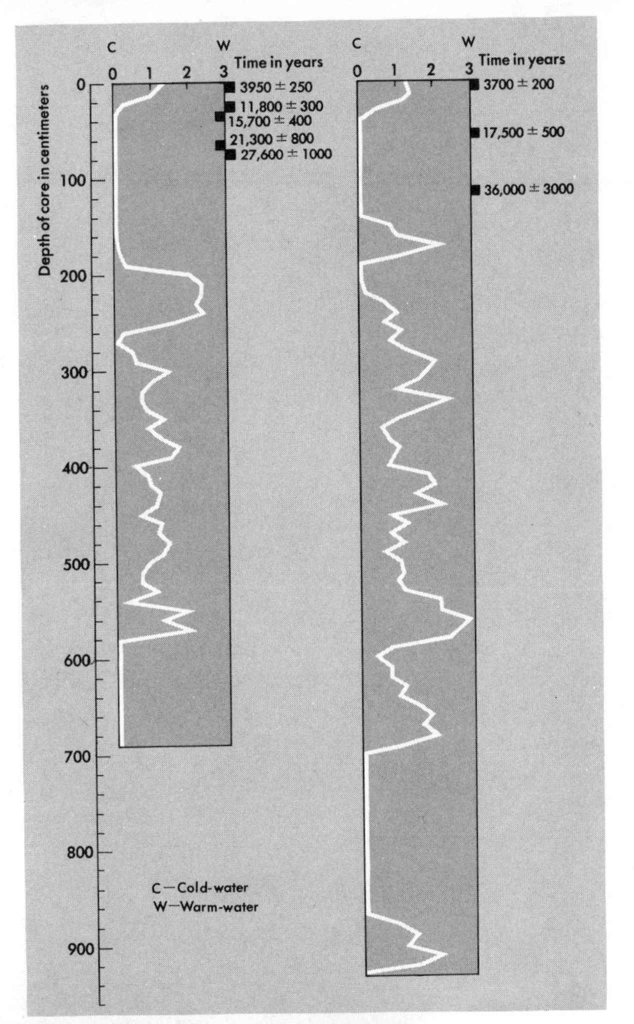

FIG. 6-15 Variations in abundance of the foraminiferan *Globorotalia menardii* in two cores from the Caribbean. The curves record changing ratios of *G. menardii* to weight of sediment coarser than 74 micrometers; low ratios indicate colder water temperatures, higher ratios, warmer water (ratios shown at top of columns). Cold intervals do not occur at the same exact depth in the cores owing to relatively higher rates of sedimentation in the core at right than in that at left. Dates to right of each core were obtained by C^{14} method; note that since 10 to 15 thousand years ago the Caribbean has become warmer. Photograph (below) shows several specimens of *Globorotalia menardii* enlarged about 22 diameters. (Drawing from Ericson and others, 1971; photo from Ericson and Wollin, 1964.)

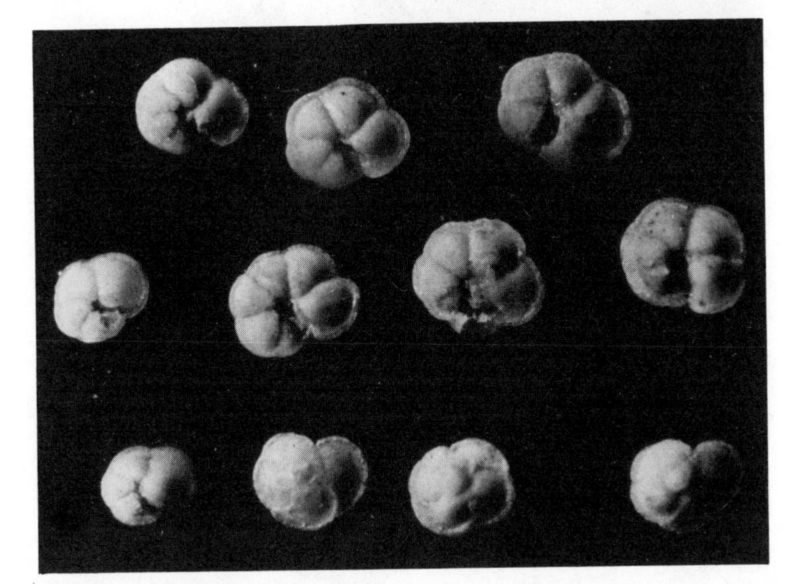

Besides *G. menardii*, some 15 to 20 other planktonic foraminiferal species and subspecies are used to define the deep-sea Pleistocene and Holocene stratigraphy of the North Atlantic. But, once having established the sequence of foraminiferal assemblages in any one core, it is necessary to correlate it with the sequences established in other cores. The reason for this is that none of the cores contains a continuous record of the Pleistocene Epoch. These hiatuses are due to submarine slumping or submarine erosion by turbidity currents.

One useful method for correlating cores is based on a surprising phenomenon: Some foraminiferal species apparently change their direction of coiling with changes in water temperature. For example, *Globigerina pachyderma*, the only planktonic foraminiferan living in the Arctic Ocean, coils to the left. But farther south, in subarctic waters of the North Atlantic it coils to the right. In this species, at least, coiling direction is dependent on water temperature (Fig. 6-16). A second species of foraminiferan, *Globorotalia truncatulinoides*, also changes its direction of coiling from cold to warm surface waters, except this species coils to the left in warmer waters and to the right in colder waters, just the converse of *Globigerina pachyderma*. Why these coiling directions occur is not known; but nevertheless it has been shown empirically that they are indeed correlated with temperature changes.

FIG. 6-16 Correlation between coiling direction of *Globigerina pachyderma* and surface water temperature. This foraminiferan coils to the left in colder waters and to the right in warmer waters. The boundary between the two different populations parallels the 7.2°C April isotherm. (From Ericson and Wollin, 1964.)

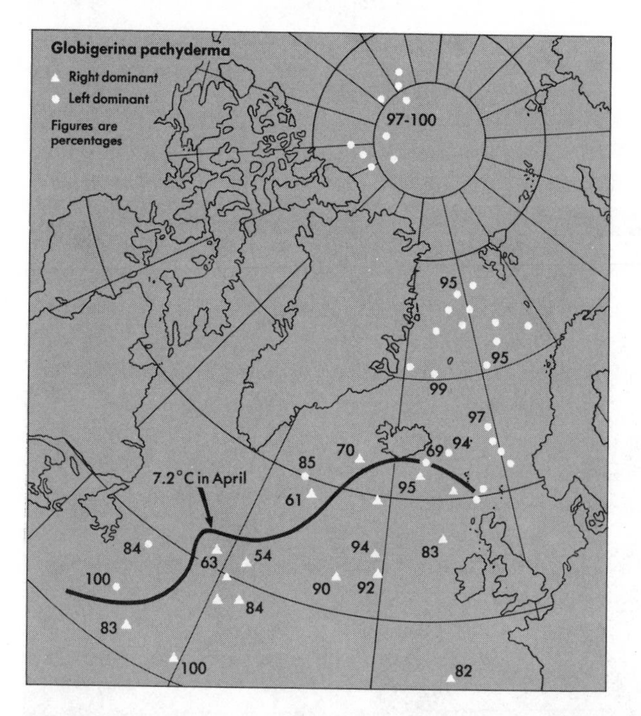

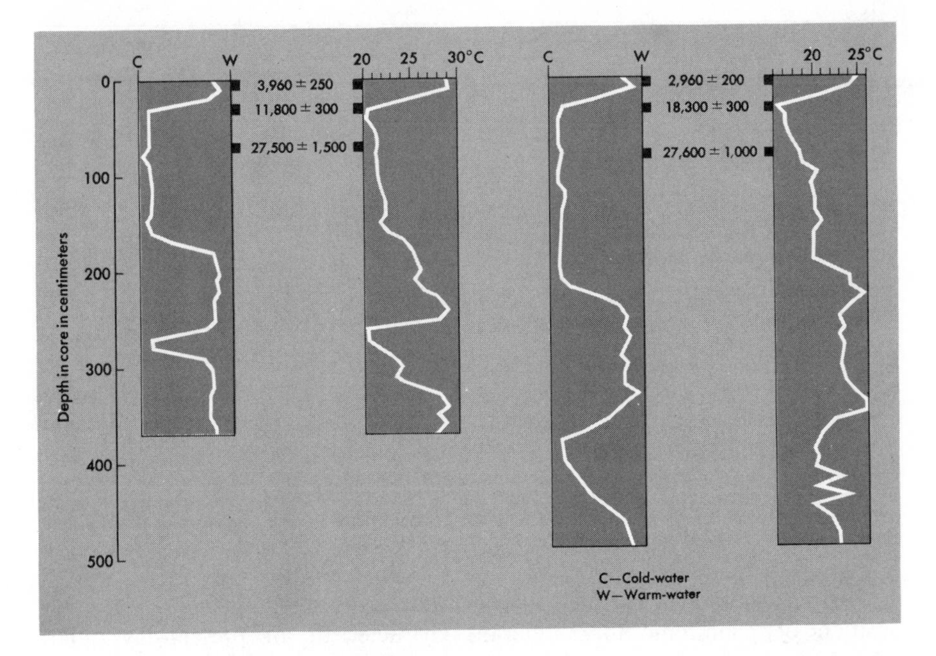

FIG. 6-17 Two cores from the North Atlantic Ocean showing close agreement in each of the climatic curves based on foraminferal assemblages and oxygen isotope ratios; absolute ages based on C_{14} measurements are also given. Note that sediment accumulation rates in the upper part of each core are similar, but that in the lower portions (below 200 centimeters) the core on the right indicates somewhat more rapid accumulation of sediment.

Further support for the conclusion that relative abundance of *Globorotalia menardii* and coiling direction of *Globorotalia truncatulinoides* mark changes in Pleistocene water temperatures of the North Atlantic is provided by oxygen isotopes. It has been observed that the relative amount in sea water of the two isotopes of oxygen, O^{18} and O^{16}, varies with water temperature (refer back to Chapter 5). Planktonic foraminiferans, in secreting their $CaCO_3$ shells, use the oxygen isotopes in the same proportion as in the surrounding sea water. Thus, shells secreted in colder water have a relatively higher ratio of O^{18} to O^{16} than do tests secreted by these same organisms in warmer water. The occurrence of these isotopic relationships in foraminiferans in North Atlantic cores agrees with the warm–cold intervals determined by *Globorotalia menardii* abundance ratios and by reversals in coiling directions of *Globorotalia truncatulinoides* and *Globigerina pachyderma* (Fig. 6-17).

After estimating ocean temperatures backward in time from single North Atlantic cores, it is possible to analyze the geographic patterns of temperature change over the whole ocean basin. For example, the CLIMAP Project (Climate: Long range investigation, mapping, and prediction) seeks to examine the history of global climate change over the last several million years as recorded

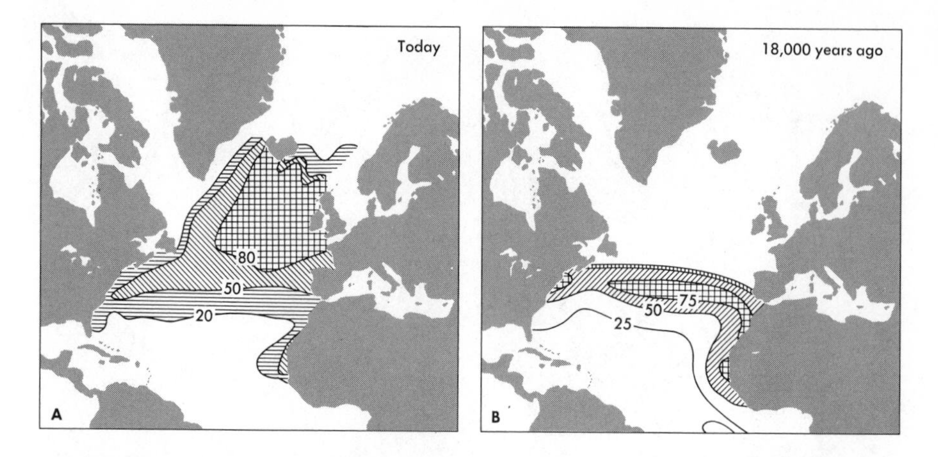

FIG. 6-18 Relative abundances of recent foraminiferans characterizing a subpolar assemblage (A) and their distribution some 18,000 years ago, during the peak of the last ice age (B). Note the northward migration of the subpolar assemblage, no doubt due to the warming of the northern polar seas following glaciation. (After Kipp and others, 1976; McIntyre and others, 1976.)

in deep-sea sediments. Part of this project involves mapping the species composition of planktonic foraminiferans characteristic of present-day polar, subpolar, subtropical, and tropical water masses of the North Atlantic based on their occurrences at the tops of deep-sea cores. Knowing then which assemblage of species lives in which water mass today, we can measure their relative abundances downward in these same cores to see how the position of the water masses may have varied in the past. As shown in Fig. 6-18, the position of the subpolar assemblage of foraminiferans was much further south during the peak of the last ice age, some 18,000 years ago.

Quantitative Environmental Changes

What we have just described are qualitative changes in environment, namely temperature, during late Pleistocene time in the North Atlantic. These changes in temperature are, of course, related to the waxing and waning of glaciation in the Northern Hemisphere as seen on the North American and European continents. The crucial, implicit assumption made in these qualitative changes in Late Pleistocene paleoecology is that present-day marine species of plankton and their immediate Pleistocene ancestors respond identically to the marine ecology. That is, we have assumed that modern-day *Globorotalia menardii* and *Globigerina pachyderma*, for example, respond to changes in water mass temperature just the way that the Pleistocene members of these same species did during the last several hundred thousand years of the Late Pleistocene Epoch. This assumption appears warranted, because the climatic results based upon these species are internally consistent and agree with other independent climatic indicators such as oxygen-isotope ratios.

More recently, paleoecologists have taken this assumption to its logical conclusion by *quantitatively* measuring past changes in North Atlantic water masses using temperature and salinity tolerances of many present-day species of marine plankton. This quantitative methodology includes the following procedures: determining the present-day composition of plankton accumulating on the sea floor from a large number of locations in widely differing latitudes and longitudes; obtaining average winter and summer temperatures as well as average salinity of the surface waters for each of these locations; and then

FIG. 6-19 Quantitative estimates of temperature and salinity of Caribbean Sea water for the last one-half million years based on the relative abundances of planktonic foraminiferans whose present-day temperature and salinity preferences are known. These estimates, based on a single core, are independently supported by oxygen isotope data from the tests of the same foraminiferans (refer back to Chapter 5). (After Imbrie and Kipp, 1971.)

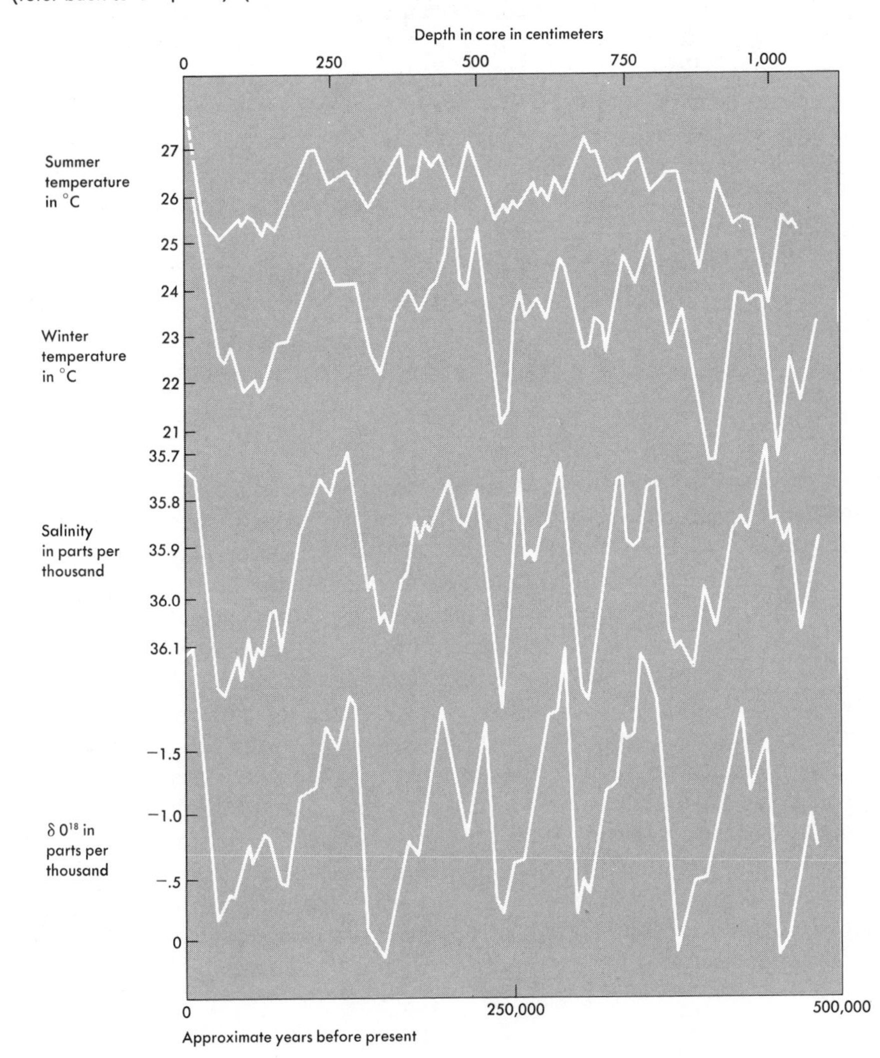

determining how the plankton changes in composition going downward in these same cores. The changes in planktonic composition backward in time (going down the core) presumably reflect corresponding changes in temperature and salinity of the overlying water mass, assuming that the plankton are ecologically related to these environmental parameters.

This method has been applied to a long core from the Caribbean Sea, and quantitative systematic variations in temperature and salinity have been recorded for almost one-half million years at this one location (Fig. 6-19). Oxygen-isotope data from the calcareous tests of these same microfossils strongly support the temperature-salinity variations based solely on the faunal composition of the different levels in the core. As more and more such cores are studied, it becomes possible to contour temperatures and salinity values of the North Atlantic—and other oceans as well, as the methodology is extended—at specific times in the past (Fig. 6-20). One exciting outcome of this research is the recognition of climatic cycles over the last half-million years, and of how those cycles might continue in the future. Here we see not only an application of the present to explain the past (using modern plankton to determine past temperatures and salinity of the oceans), but also using the past to predict the future. In particular, one theory holds that past glacial ages have been due to small but significant changes in the Earth's orbital geometry thereby causing differences in solar insolation. Furthermore, these changes are likely to continue, leading to extensive Northern Hemisphere cooling and glaciation over the next 20,000 years.

SUMMARY

The distribution and abundance of fossils can be explained in terms of the original ecologic conditions of their ancient environments. In the case of older, mostly extinct biotas, we must rely heavily on the paleoecological evidence provided by the enclosing sedimentary rocks. Environmental stratigraphy can therefore provide ecologic understanding and insight, independent of the fossils themselves. Thus, the depositional environments for early Devonian marine limestones and early Permian nonmarine shales can be inferred from the lithofacies characteristics of these rocks. The knowledge of modern shallow carbonate and deltaic environments, respectively, can also shed light on the origin of these rocks. Having established the ancient habitats, it is then possible to relate specific fossil occurrences to them, in terms of the probable ways of life of the associated organisms, based on comparison with approximately similar adaptive types. Knowing the biologic needs and ecologic preferences of modern organisms like mat-building blue-green algae, epifaunal suspension feeding brachiopods, burrowing lungfish, or predatory lizard-like reptiles, we can reconstruct the ancient communities and ecosystems, at least in part.

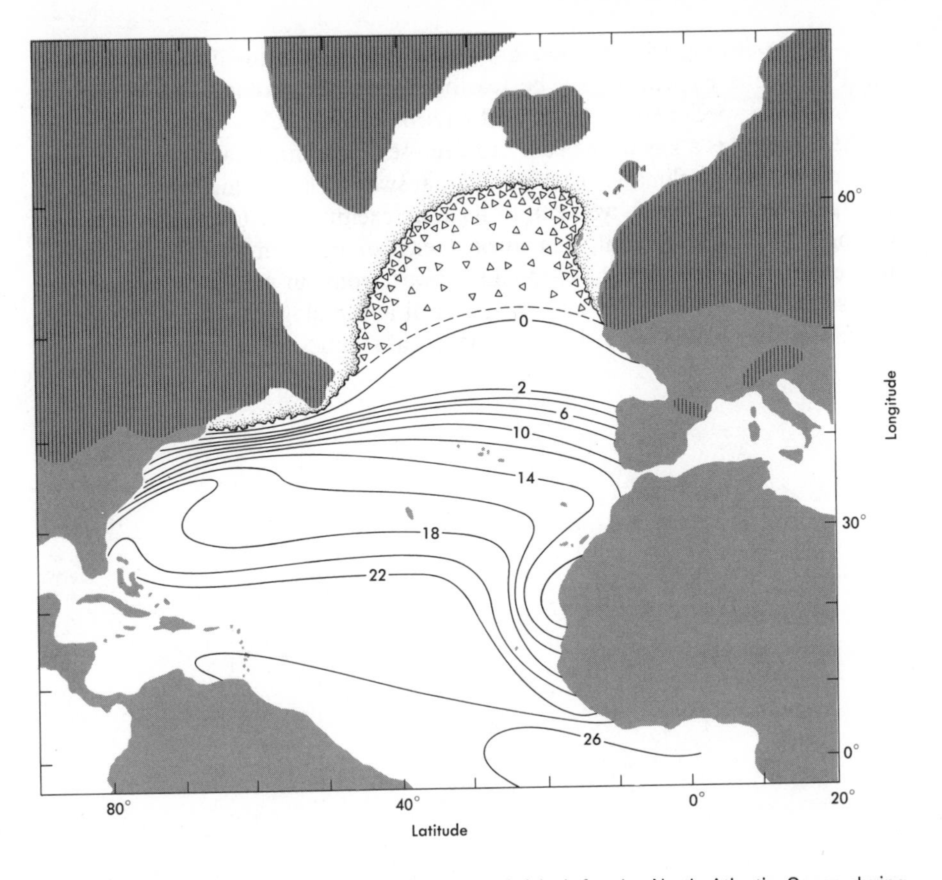

FIG. 6-20 Winter surface temperatures (degrees Celsius) for the North Atlantic Ocean during the last maximum glaciation, some 18,000 years ago. This map is based upon planktonic foraminiferans and coccolithophorids found in North Atlantic cores at levels of this age, and assumes that these planktonic microfossils had the same ecologic requirements then as now. Compare this map with the distribution of the subpolar foraminiferan assemblage shown in Fig. 6-18 (B). Continental glaciers are shown by hachures, pack ice by stippling, and loose pack ice by triangles. Glacial shorelines are drawn using present bathymetry lowered by 100 meters. (After McIntyre and others, 1976.)

After establishing the nature of specific communities at one particular point in time, we may wish to see how these communities might have changed through time. While nearshore, carbonate marine communities of middle Paleozoic time appear to have been hardly changed at all, late Paleozoic and early Mesozoic terrestrial communities changed considerably. The former example comes somewhat after the rapid diversification of marine life in the early Paleozoic Era and may therefore represent a period of stable equilibrium, but the latter example comes during the expansion and diversification of terrestrial life before such an equilibrium had yet been established.

When dealing with the much younger part of the fossil record, such as the Pleistocene Epoch, it may be feasible to use the ecology and biology of extant, living species to interpret past environments. Such a literal application of the "present is a key to the past" to deep-sea environmental conditions, and, in turn, to global climate, may hold the answer to predicting future environmental conditions. Thus, we seem to be at an exciting threshold in the study of geologic history in general, and ancient environments in particular, namely, where we can look both backward and forward from our present instant in time and see the broad sweep of events—physical and biological—unfolding across the face of our planet, as a single, unified, and seamless time warp.

suggestions for further reading

CHAPTER 1

GEOLOGIC ENVIRONMENTS

AGER, D. V., 1963, *Principles of Paleoecology*, New York: McGraw-Hill Book Company.

CLOUD, P. E., JR., 1959, Paleoecology—retrospect and prospect, *J. Paleontol.*, v. 33, p. 926–962.

CRAIG, G. Y., 1966, Concepts in paleoecology, *Earth Science Reviews*, v. 2, p. 127–155.

IMBRIE, J. and NEWELL, N. D. (editors), 1964, *Approaches to Paleoecology*, New York: John Wiley and Sons, Inc.

SIMPSON, G. G., 1963, Historical science, *The Fabric of Geology*, C. C. Albritton, Jr., editor, Reading, Mass.: Addison-Wesley Publishing Co., Inc., p. 24–48.

CHAPTER 2

SEDIMENTS AND ENVIRONMENTS

ALLEN, J. R. L., 1970, *Physical Processes of Sedimentation*, New York: American Elsevier Publ. Co.

BATHURST, R. G. C., 1971, *Carbonate Sediments and Their Diagenesis*, New York: American Elsevier Publ. Co.

BLATT, H., MIDDLETON, G., and MURRAY, R., 1972, *Origin of Sedimentary Rocks*, Englewood Cliffs, N.J.: Prentice-Hall, Inc.

MATTHEWS, R. K., 1974, *Dynamic Stratigraphy*, Englewood Cliffs, N.J.: Prentice-Hall, Inc.

SELLEY, R. C., 1976, *Introduction to Sedimentology*, New York: Academic Press, Inc.

REINECK, H. and SINGH, I. B., 1973, *Depositional Sedimentary Environments—with Reference to Terrigenous Clastics*. Berlin: Springer-Verlag.

CHAPTER 3

ORGANISMS AND ENVIRONMENTS

McCONNAUGHEY, B. H., 1974, *Introduction to Marine Biology*, St. Louis: The C. V. Mosby Company.

HEDGPETH, J. W., and LADD, H. (editors), 1957, Treatise on Marine Ecology and Paleoecology, *Geol. Soc. Am., Mem.* **67**, 2 vols.

MacGINITIE, G. E. and MacGINITIE, N., 1949, *Natural History of Marine Animals*, New York: McGraw-Hill Book Company.

MOORE, H. B., 1958, *Marine Ecology*, New York: John Wiley and Sons, Inc.

SCHÄFER, W., 1972, *Ecology and Palaeoecology of Marine Environments*, Chicago: The University of Chicago Press.

SIMPSON, G. G. and BECK, W., 1965, *Life: An Introduction to Biology*, 2nd edition, New York: Harcourt, Brace and World.

CHAPTER 4

TAPHONOMY

BEHRENSMEYER, A. K., 1975, The taphonomy and paleoecology of Plio-Pleistocene vertebrate assemblages east of Lake Rudolf, Kenya, *Mus. Comp. Zool. Bull.*, v. 146, p. 473–578.

KONIZESKI, R. L., 1957, Paleoecology of the Middle Pliocene Deer Lodge local fauna, Western Montana, *Geol. Soc. Am. Bull.*, v. 68, p. 131–150.

LAWRENCE, D. R., 1968, Taphonomy and information losses in fossil communities, *Geol. Soc. Am. Bull.*, v. 79, p. 1315–1330.

VOORHIES, M., 1969, *Taphonomy and Population Dynamics of an early Pliocene Vertebrate Fauna, Knox Co., Nebraska*, Contrib. to Geol. Univ. Wyoming, Spec. Paper 1.

WARME, J., 1971, Paleoecological aspects of a modern coastal lagoon, *Univ. Calif. Publ. Geol. Sci.*, v. 87, p. 1–112.

CHAPTER 5

ENVIRONMENTAL ANALYSIS

BROECKER, W. S., 1974, *Chemical Oceanography,* New York: Harcourt Brace Jovanovich, Inc.

FRIEDMAN, G. (editor), 1969, *Depositional Environments in Carbonate Rocks,* Soc. Econ. Paleontologists and Mineralogists, Spec. Publ. 14.

HECKER, R. F., 1965, *Introduction to Paleoecology,* New York: American Elsevier Publ. Co.

RIGBY, J. K. and HAMBLIN, W. K. (editors), 1972, *Recognition of Ancient Sedimentary Environments,* Soc. Econ. Paleontologists and Mineralogists, Spec. Publ. 16.

SHAW, A. B., 1964, *Time in Stratigraphy,* New York: McGraw-Hill Book Company.

CHAPTER 6

ENVIRONMENTAL SYNTHESIS

DARWIN, C., 1859, *On the Origin of Species* (facsimile of 1st edition, 1964, Cambridge, Mass.: Harvard University Press).

DOBZHANSKY, T., AYALA, F. J., STEBBINS, G. L., and VALENTINE, J., 1977, *Evolution,* San Francisco: W. H. Freeman and Company, Publishers.

OLSON, E. C., 1971, *Vertebrate Paleozoology,* New York: Wiley-Interscience.

SCHOPF, T. J. M. (editor), 1972, *Models in Paleobiology,* San Francisco: Freeman, Cooper, and Co.

SCOTT, R. W. and WEST, R. R. (editors), 1976, *Structure and Classification of Paleocommunities,* Stroudsburg, Pa.: Dowden, Hutchinson, and Ross, Inc.

VALENTINE, J. W., 1973, *Evolutionary Paleoecology of the Marine Biosphere* Englewood Cliffs, N.J.: Prentice-Hall, Inc.

credits

CHAPTER 1

Figure 1-1	Augusta, J. and Burian, Z., Prague, 1960.
1-2	Pelletier, B. R., 1958, Pocono paleocurrents in Pennsylvania and Maryland, *Geol. Soc. Am. Bull.*, v. 69, p. 1055.
1-4	Aero Photographers.
Table 1-1	Crosby, E. J., 1972, *Recognition of Ancient Sedimentary Environments*, Soc. Econ. Mineralogists Paleontologists, Spec. Publ. 16, p. 10.

CHAPTER 2

Figure 2-1	Weller, J. M., 1960, *Stratigraphic Principles and Practice*, New York: Harper and Row Publishers, Inc., p. 341.
2-2	Laporte, L. F., 1975, *Encounter with the Earth*, San Francisco: Canfield Press, p. 163.
2-3	Press, F. and Siever, R., 1974, *Earth*, San Francisco: W. H. Freeman and Company, p. 291.
2-4	Passega, R., 1957, Texture as characteristic of clastic deposition, *Am. Assoc. Petrol. Geol. Bull.*, v. 41, p. 1973.
2-5	Visher, G. S., 1969, Grain size distributions and depositional processes, *J. Sed. Petrol.*, v. 39, p. 1081.
2-6	See Fig. 2-1, p. 91.
2-7	Pettijohn, F. J. and Potter, P., 1964, *Atlas and Glossary of Primary Structures*, New York: Springer-Verlag.
2-8	Ingle, J. C., Jr., 1975, *Current Concepts of Depositional Systems with Applications for Petroleum Geology*, San Joaquin Geol. Soc., p. 2–4.
2-9	Allen, J. R. L., 1970, *Physical Processes of Sedimentation*, New York: American Elsevier Publ. Co., p. 205.
2-10	Folk, R. L. and Robles, R., 1964, Carbonate sands of Isla Perez, Alacran reef complex, Yucatan, *J. Geol.*, v. 72, p. 267.
2-11	Rhoads, D. C., Yale University.
2-12	Ginsburg, R. N., 1957, *Regional Aspects of Carbonate Deposition*, Soc. Econ. Paleontologists Mineralogists, Spec. Publ. 5, p. 82.
2-13	A and B—(see Fig. 2-12, p. 87); C—Newell, N. D. and Rigby, J. R., 1957, *Regional Aspects of Carbonate Deposition*, Soc. Econ. Paleontologists Mineralogists, Spec. Publ. 5, plate VI-2.
2-14	A, B, C—Logan, B. W., Rezak, R., and Ginsburg, R. N., 1964, Classification and environmental significance of stromatolites, *J. Geol.*, v. 72, p. 72, Plate 1A, 3C. D—Rezak, R., 1957, *Stromatolites of the Belt Series in Glacier National Park and Vicinity, Montana* U.S. Geol. Surv. Prof. Paper 294-D, Plate 21-8. E—Hofmann, H., 1969, *Attributes of Stromatolites*, Geol. Surv. Canada, Paper 69-39, p. 4.
2-15	A and B—Ginsburg, R. N. and Lowenstam, H. A., 1958, The influence of marine bottom communities on the depositional environment of sediments, *J. Geol.* v. 66, p. 31. C—Matthews, R. K., 1974, *Dynamic Stratigraphy*, Englewood Cliffs, N.J.: Prentice-Hall, Inc., p. 230.

2-16 Imbrie, J. and Buchanan, H., 1965, *Sedimentary structures in modern carbonate sands of the Bahamas*, Soc. Econ. Paleontologists Mineralogists, Spec. Publ. 12, p. 158, 167.
2-17 Dunbar, C. and Rodgers, J., 1957, *Principles of Stratigraphy*, New York: John Wiley and Sons, Inc., p. 155.
2-19 Graham, S. A., 1975, *Current Concepts of Depositional Systems with Applications to Petroleum Geology*, San Joaquin Geol. Soc., p. 0–3.
2-21 See Fig. 2-19, p. 0–5.
2-23 See Fig. 2-19; p. 0–4.
2-25 See Fig. 2-19; p. 0–4.
2-26 See Fig. 2-19; p. 0–7.
Table 2-1 See Fig. 2-2, p. 47.
2-2 McAlester, A. L., 1977, *The History of Life*, Englewood Cliffs, N.J.: Prentice-Hall, Inc., p. 53.

CHAPTER 3

Figure 3-3 Moore, H. B., 1968, *Marine Ecology*, New York: John Wiley and Sons, Inc., p. 34, 35.
3-4 See Fig. 3-3; p. 26.
3-5 Segerstråle, S. A., 1957, Treatise on Marine Ecology, *Geol. Soc. Am., Mem.* 67, v. 1, p. 777.
3-6 See Fig. 1-1; p. 321.
3-7 Kettlewell, H. B. D., Oxford University.
3-8 Purdy, E. G., 1964, Sediments as Substrates, in *Approaches to Paleoecology*, Imbrie, J. and Newell, N. D. (editors), New York: John Wiley and Sons, Inc., p. 260.
3-9 Newell, N. D., Imbrie, J., Purdy, E. G., and Thurber, D. L., 1959, Organism communities and bottom facies, Great Bahama Bank, *Am. Mus. Nat. Hist. Bull.*, v. 117, p. 202.
3-10 See Fig. 3-9, p. 199, 201.
3-11 Rhoads, D. C. and Berner, R. A., 1968, *Animal-Sediment Relationships and Early Sediment Diagenesis in Long Island Sound*, New England Intercollegiate Geol. Conf. Guidebook, Trip. E-1, p. 9.
3-12 See Fig. 1-1; p. 646.
3-13 Goreau, T. F., 1961, Problems of growth and calcium deposition in reef corals, *Endeavour*, v. 20, p. 38.
3-14 Hedgpeth, J., 1957, Treatise on Marine Ecology, *Geol. Soc. America, Mem.* 67, v. 1, p. 37.
3-15 See Fig. 1-1, p. 643.
3-18 Fischer, A. G., 1960, Latitudinal variations in organic diversity, *Evolution*, v. 14, p. 69, 70, 71.
3-19 Sanders, H., 1968, Marine benthic diversity: a comparative study, *Amer. Naturalist*, v. 102, p. 250.
Table 3-1 Simpson, G. G., 1953, *Evolution and Geography*, Oregon State System of Higher Education, Condon Lectures, p. 27.

CHAPTER 4

Figure 4-1 Raup, D., 1976, Species diversity in the Phanerozoic, *Paleobiology*, v. 2, p. 286, 291.
4-2 Behrensmeyer, A. K., 1975, The taphonomy and paleoecology of Plio-Pleistocene vertebrate assemblages east of Lake Rudolf, Kenya, *Mus. Comp. Zool. Bull.*, v. 146, p. 493.
4-3 Konizeski, R. L., 1957, Paleoecology of the Middle Pliocene Deer Lodge local fauna, Western Montana, *Geol. Soc. Am. Bull.*, v. 68, p. 147.
Table 4-2 Voorhies, M. R., 1969, *Taphonomy and Population Dynamics of an Early Pliocene Vertebrate Fauna, Knox Co., Nebraska*, Contrib. to Geology, Univ. Wyoming, Spec. Paper 1, p. 69.
4-3 Warme, J., 1971, Paleoecological aspects of a modern coastal lagoon, *Univ. Calif. Publ. Geol. Sci.*, v. 87, p. 94.

4-4 Lawrence, D. R., 1968, Taphonomy and information losses in fossil communities, *Geol. Soc. Am. Bull.*, v. 79, p. 1323.
4-5 See Table 4-4, p. 1325.
4-6 See Table 4-4, p. 1326.
4-7 See Fig. 4-3, p. 135.
4-8 See Fig. 4-3, p. 148.

CHAPTER 5

Figure 5-2 Irwin, M. L., 1965, General theory of epeiric, clear water sedimentation, *Am. Assoc. Petrol. Geol. Bull.*, v. 49, p. 450.
5-3 See Fig. 5-2; p. 446.
5-4 A—Walker, K. R. and Bambach, R. K., 1974, Feeding by benthic invertebrates: classification and terminology of paleoecological analysis, *Lethaia*, v. 7, p. 67–78. B—See Fig. 3-8; p. 241, 242.
5-5 Walker, K. R. and Bambach, R. K., 1974, *Principles of Benthic Community Analysis*, Sedimenta IV, Comp. Sedimentology Laboratory, Univ. Miami, p. 2–7.
5-6 La Roque, A., 1949, Post-Pleistocene connection between James Bay and the Gulf of St. Lawrence, *Geol. Soc. Am. Bull.*, v. 60, p. 375.
5-7 Eicher, D. L., 1969, Paleobathymetry of Cretaceous Greenhorn Sea in Eastern Colorado, *Am. Assoc. Petrol. Geol.*, v. 53, p. 1078.
5-8 See Fig. 5-7, p. 1079.
5-9 Heckel, P., 1972, *Recognition of Ancient Sedimentary Environments*, Soc. Econ. Paleontologists and Mineralogists, Spec. Publ. 16, p. 244.
5-10 Lowenstam, H. A., 1954, Factors affecting the aragonite: calcite ratios in carbonate-secreting organisms, *J. Geol.*, v. 62, p. 285.
5-11 Chave, K., 1954, Aspects of the biogeochemistry of magnesium. 1. Calcareous marine organisms, *J. Geol.*, v. 62, p. 277, 281.
5-12 Kay, M. and Colbert, E., 1965, *Stratigraphy and Life History*, New York: John Wiley and Sons, Inc., p. 393.
5-13 Broecker, W. S., and Van Donk, J., 1970, Insolation changes, ice volume, and the ^{18}O record in deep-sea cores. *Rev. Geophys. Space Phys.*, v. 8, p. 172.
5-14 Garrels, R. M. and Mackenzie, F. T., 1971, *Evolution of Sedimentary Rocks*, New York: W. W. Norton and Co., Inc., p. 91.
5-15 Fritz, P., Anderson, T. W., and Lewis, C. F. M. 1975, Late-Quaternary climatic trends and history of Lake Erie from stable isotope studies, *Science*, v. 190, p. 268.

Table 5-1 Laporte, L., 1967, Carbonate deposition near mean sea-level and resultant facies mosaic: Manlius formation (Lower Devonian) of New York State, *Am. Assoc. Petrol. Geol. Bull.*, v. 51, p. 90.
5-2 Goldring, W. W., 1922, The Champlain Sea, *N.Y. State Mus. Sci. Bull.*, v. 239, p. 164.

CHAPTER 6

Figure 6-1 Laporte, L., 1969, *Depositional environments in carbonate rocks*, Soc. Econ. Paleontologists and Mineralogists, Spec. Publ. 14, p. 101.
6-2 See Fig. 6-1; p. 103.
6-3 See Fig. 6-1; p. 104.
6-4 See Fig. 6-1; p. 116.
6-5 Walker, K. R. and Laporte, L., 1970, Congruent fossil communities from Ordovician and Devonian carbonates of New York, *J. Paleontol.*, v. 44 p. 938, 939.
6-6 Olson, E. C., 1971, *Vertebrate Paleozoology*, New York: Wiley-Interscience, p. 644.
6-7 Olson, E. C., UCLA.

6-8 Olson, E. C., 1958, Fauna of the Vale and Choza: 14, Summary, Review, and Integration of the Geology and the Faunas, *Fieldiana-Geology*, v. 10, p. 437.
6-9 See Fig. 6-6; p. 638.
6-10 Olson, E. C., 1952, The Evolution of a Permian Vertebrate Chronofauna, *Evolution*, v. 6 p. 181–196.
and
Olson, E. C., 1966, Community evolution and the origin of mammals, *Evolution*, v. 47, p. 291–302.
6-11 Olson, E. C., 1961, *The Evolution of Lower and Non-Specialized Mammals I*, Koninklijke Vlaamse Academie voor Wetenschappen, Latteren en Schone Kunsten van Belgie, p. 101.
6-12 See Fig. 6-11, p. 113.
6-13 Ericson, P. and Wollin, G., 1964, *The Deep and the Past*, New York: Alfred A. Knopf, Inc., p. 45.
6-14 See Fig. 6-13; Plates IV, V.
6-15 Ericson, D. and Wollin, G., 1961, Atlantic deep-sea cores, *Geol. Soc. Am. Bull.*, v. 72, p. 265.
6-16 See Fig. 6-13, p. 89.
6-17 See Fig. 6-15, p. 277.
6-18 A–Kipp, N., 1976, Investigation of Late Quaternary Paleoceanography and Paleoclimatology, *Geol. Soc. Am., Mem.* 145, p. 25.
B—McIntyre, A., and others, 1976, Investigation of Late Quaternary Paleoceanography and Paleoclimatology, *Geol. Soc. Am., Mem.* 145, p. 51.
6-19 Imbrie, J. and Kipp, N. G., 1971, *The Late Cenozoic Glacial Ages*, New Haven: Yale University Press, p. 118.
6-20 See Fig. 6-18B, p. 59.
Table 6-1 See Fig. 6-1, p. 105.
6-2 See Fig. 6-5, p. 935.
6-3 See Fig. 6-6, p. 649.

index

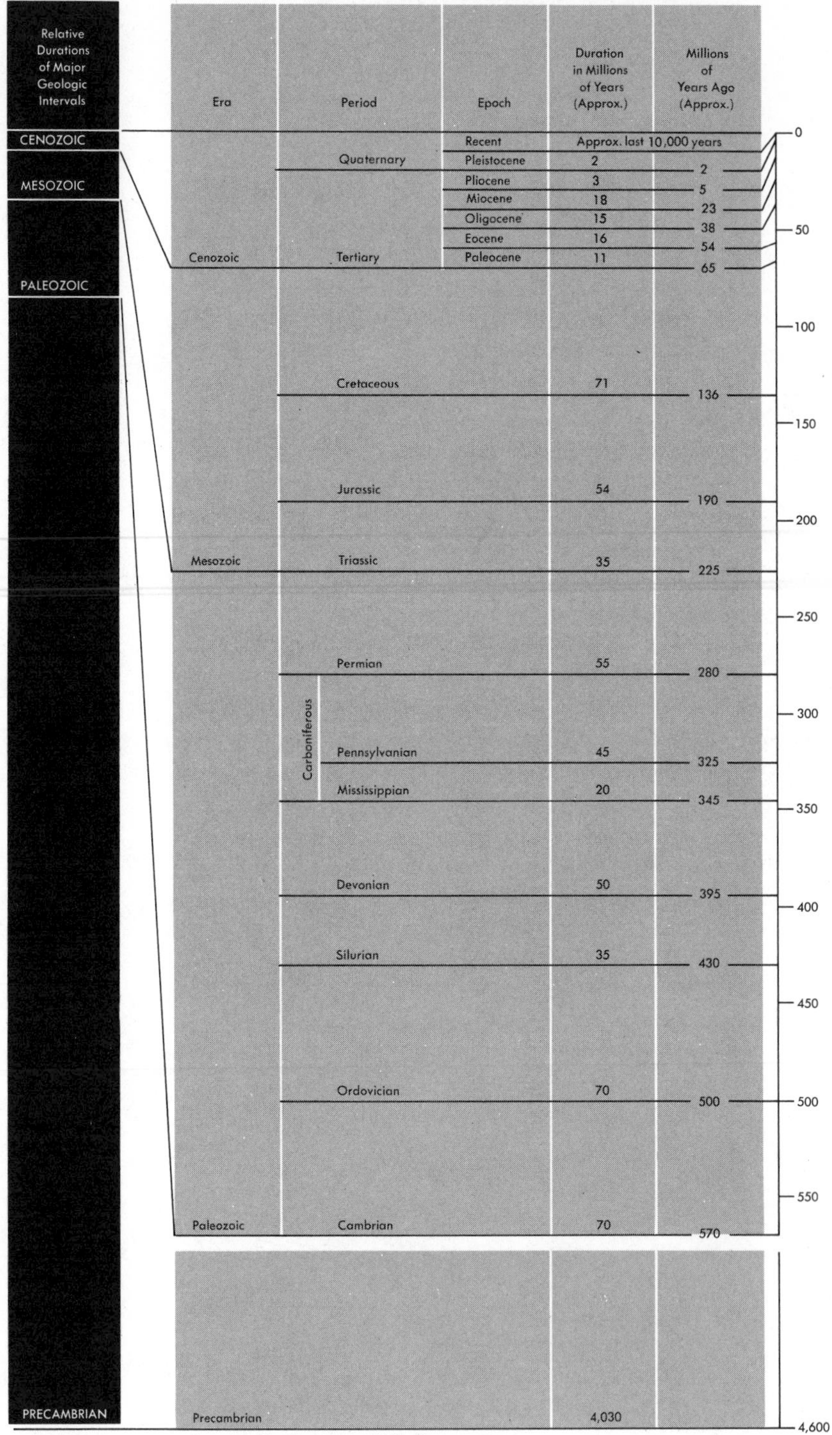

Era	Period	Epoch	Duration in Millions of Years (Approx.)	Millions of Years Ago (Approx.)
Cenozoic	Quaternary	Recent	Approx. last 10,000 years	
		Pleistocene	2	2
	Tertiary	Pliocene	3	5
		Miocene	18	23
		Oligocene	15	38
		Eocene	16	54
		Paleocene	11	65
Mesozoic	Cretaceous		71	136
	Jurassic		54	190
	Triassic		35	225
Paleozoic	Permian		55	280
	Carboniferous: Pennsylvanian		45	325
	Carboniferous: Mississippian		20	345
	Devonian		50	395
	Silurian		35	430
	Ordovician		70	500
	Cambrian		70	570
Precambrian			4,030	

Formation of Earth's crust about 4,600 million years ago